Internationally renowned author and horti-culturalist Dr W. E. Shewell-Cooper has written more than 60 best-selling books on gardening, beginning with *The Garden* in 1932, and contributes to 40 newspapers and journals. He frequently appears on radio and television and travels all over the world to lecture and advise. He is best known as the champion of compost growing, an organic gardening method which he first re-searched at the Thaxted Horticultural College, Essex, where he was Principal for ten years. Evidence of the success of this method – described in detail in this book – can now be seen at Dr Shewell-Cooper's famous experimental gardens at Arkley Manor, Arkley, Hertfordshire.

Recognition has come from many countries, including France (Commandeur du Mérite Agricole), Austria (Fellow and Doctor of the Vienna Horticulture College), and Italy (Cavaliere al Merito). Dr Shewell-Cooper is Director of the International Horticultural Advisory Bureau and founder and Hon. Director of The Good Gardeners' Association.

W. E. Shewell-Cooper

MBE, NDH, FLS, FRSL, MRST, Dip Hort (Wye), D Litt

Soil, Humus and Health

An Organic Guide

MAYFLOWER
GRANADA PUBLISHING
London Toronto Sydney New York

Published by Granada Publishing Limited
in Mayflower Books 1978

ISBN 0 583 12796 7

First published in Great Britain by
David & Charles Ltd 1975
Copyright © W. E. Shewell-Cooper 1975

Granada Publishing Limited
Frogmore, St Albans, Herts AL2 2NF
and
3 Upper James Street, London, W1R 4BP
1221 Avenue of the Americas, New York, NY 10020, USA
117 York Street, Sydney, NSW 2000, Australia
100 Skyway Avenue, Toronto, Ontario, Canada M9W 3A6
Trio City, Coventry Street, Johannesburg 2001, South Africa
CML Centre, Queen & Wyndham, Auckland, New Zealand

Made and printed by
Richard Clay (The Chaucer Press) Ltd
Bungay, Suffolk
Set in Linotype Times

Enneus (239-169 BC):
   And Earth herself who bestowed the body takes it back
   and wastes not a whit.

Cicero (106-43 BC):
   What I enjoy is not the fruit alone. I also enjoy the soil
   itself and its power.

Alexander Pope (1688-1744):
   Nothing is foreign; parts relate to whole;
   One all-extending, all-preserving soul
   Connects each being, greatest with the least;
   Made beast in aid of Man, and Man of beast;
   All served, all serving; nothing stands alone;
   The chain holds on, and where it ends unknown.

William Temple (Archbishop of Canterbury, 1942–5, and
cousin of the author):
   The treatment of earth by Man the exploiter is not only
   improvident but sacrilegious. We are not likely to correct
   our hideous mistakes in this realm unless we can recover
   the mystical sense of oneness with Nature. I labour this
   precisely because many people think it fantastic; I think
   it fundamental to sanity.

Lady Eve Balfour in *Mother Earth* (1946):
   Disorder and chaos are not natural phenomena. Left to
   herself, Nature always produces order. It is Man who
   causes chaos by his persistent attempts to resist or ignore
   natural laws, an attempt doomed to failure from the start.

CONTENTS

Units of Measurement

All units of measurement in this volume are given in the SI (metric) system. The following list shows the conversions between British units and the SI system.

Distance
1 inch	=	25.4 millimetres (mm)
1 foot	=	0.3048 metres (m)
1 yard	=	0.9144 metres (m)

Area
1 square foot	=	0.0929 square metres (m²)
1 square yard	=	0.836 square metres (m²)
1 acre	=	0.40 hectares

Volume
1 imperial pint	=	0.57 litres
1 imperial gallon	=	4.546 litres
1 bushel	=	36.37 litres

Weight
1 ounce	=	28.35 grammes (g)
1 pound	=	0.454 kilogrammes (kg)
1 hundredweight	=	50.8 kilogrammes (kg)
1 ton	=	1016 kilogrammes = 1 tonne

PREFACE

For some thirty-five years now I have had the privilege of teaching students at horticultural colleges, of lecturing to allotment societies and the like, of advising units in the army, of helping in school gardens, and teaching in evening institutes, as well as acting as garden editor to a number of newspapers and journals. This continual personal contact has made me realize how interested people are in the soil, and the best ways of feeding it.

I am a Foundation Member of The Soil Association, and in 1964 founded the Good Gardeners' Association of which I am chairman. This association has done much to foster interest in our heritage, 'the soil', emphasizing the necessity of conserving all vegetable waste and returning it to the soil where it belongs.

Some scientists may claim that by soil analysis they can produce the answer for constant richness and fertility, but there is always the problem of that mystical substance, humus, what it does and what it is. Do we really know whether acidity in soil is the same as soil bitterness or sourness? Are we quite satisfied that lime is always the answer? As a gardener I am learning day by day, and even this book does not pretend to be the final answer to all soil problems. Here I attempt to put into straight-forward language important facts which should help gardeners, both professional and amateur, to grow better fruit, better vegetables and better flowers.

The Good Gardeners' Association has increased its numerical strength year by year and as I write we have enrolled our 5900th Fellow. This book is considered the handbook of the association, for here, correct composting is taught and no purely chemical fertilizers are ever advised.

I should like to pay my tribute to Lady Eve Balfour who first envisaged the founding of the Soil Association. She has helped me again and again with her lectures and advice. I must pay a tribute also to the late Sir Albert Howard for being such a doughty crusader. I would express my indebtedness also to the late Mr F. Newman Turner, NDA, NDD, who worked with me for many years at our former research station in Thaxted.

My thanks are due too to Mrs S. V. Walker for typing the script, and to my friend Mr J. Frank Milner for reading the original proofs.

W. E. Shewell-Cooper
Hon. Director

The Good Gardeners' Association
Arkley Manor
Near Barnet
Hertfordshire

CHAPTER 1

The Introduction

It was when lecturing to hundreds of army units during the war and afterwards that I discovered how few town-dwellers realized how much human beings depend on the soil. It was my job and privilege as a senior officer in the army in the early 1940s to ensure that the land taken over by the government was cultivated to its fullest extent. After the Allies captured Germany the 21st Army Group could not live off the land, for the Germans were starving, and once again I was sent on a mission to see that the ground was properly cultivated and the maximum crops produced.

All this taught me the very great need for everyone to realize that we depend on the soil for our being. For people who live in towns, it is easy to forget the sowing, manuring and cultivation involved in producing food before it reaches the cities.

Country folk are apt to laugh when I tell them that town-dwellers are losing their connection with food production. I was showing some men from the East End of London around our gardens some time ago, when I was asked the name of a vegetable. It was a row of broad beans! Only two men in a party of thirty or more had ever seen the broad bean growing.

It is not only food that comes from the soil. The basic material for cotton shirts comes from the cotton plant; artificial silk for women's underwear is from the wood of a tree, while the true silk garment is produced from the silkworm who lives on the leaves; the leather with which our shoes are made is from the skins of animals which feed on the soil; most of the furniture in my office, for instance, is made from a tree which once grew in the soil.

Some have argued that it is possible to grow certain plants in water, but hydroponics (as this culture has come to be

13

called can never keep the world supplied with all its needs. Furthermore, it is doubted whether any food produced in this way would truly be of the same value as food produced in soil that has been regularly enriched with humus. Such culture, therefore, must be regarded as but an interesting hobby, or an area for scientific experiment.

The tragedy of the great dust-bowl in the 1930s focused the attention of all gardeners and agriculturists on the need for organic soil feeding. Some 100 000 farmers in the Middle West of America were rendered homeless by soil ruination, caused by continual ploughing up of the land, the use of quantities of artificial fertilizers and the burning of straw which should have been composted and ploughed back into the soil. The resulting sandy dust could not hold moisture, and the dust storms gathered in strength. Government experts have estimated that one-fifth of the farming land in the USA has been made infertile in this way.

US officials have said in their reports that certain farming practices in Utah, Arizona, and in other parts of the Middle West, have ruined in fifty years land which took thousands of years to get into the right condition. But it is not only in the United States that these things are happening. I myself have seen serious soil erosion in parts of Africa and Cyprus, and reports come in of similar troubles in New Zealand and Australia. It is by no means unknown in numerous areas of India, and the latest Indian government is showing concern about the position.

Mr F. A. Secrett, CBE, VMH, who was Honorary Horticultural Adviser to Britain's Ministry of Agriculture in the 1939-45 war, said when lecturing to the Royal Society of Arts:

Some scientific workers consider that in time inorganic manures may be used in place of organic animal manure, but those engaged in practical work realize that this is a dangerous fallacy ... In limiting the use of humus, has thought been given to the benefits derived from the use of organic animal manures in assisting the aeration of

14

the land and in supplying the necessary mechanical action? The country has seriously impoverished some of the best soils in England. The country in fact has been prodigal in its use of artificial manures, and its prodigality has led to want. Bedfordshire is a striking example. In past years the land in this county was heavily manured, but owing to the shortage of stable manure, artificials were resorted to, with the result that the land is now reverting to desert conditions.

Note the words 'reverting to desert conditions'. I have seen this in various parts of Britain too, such as in Norfolk, parts of Suffolk, parts of Essex and several other counties. In Lincolnshire, for instance, the blowing away of the surface soil with the seeds that had been sown, because there was not sufficient humus in the ground to hold the soil particles together, has been reported more than once. In some cases the topsoil of whole fields has gone.

There are large districts in China where market gardeners have been cultivating their land not only continuously, but intensively, for forty centuries, and yet their soil is still in a perfect condition. What the Chinese peoples do we can do. They realize the importance of passing back into the soil what has been taken out. They have taken care to conserve and prepare the organic matter from vegetable, animal and human waste by composting and are able to manure the land efficiently.

A fertile soil means healthy crops, healthy livestock, and last, but not least, healthy human beings.

Surely it is practical common sense to carry out the law of nature, ie the law of return. This implies that in the raising of crops, two complementary processes must function – growth and decay. If growth is to be speeded up, then decay must be accelerated also. The gardener who confines his attention to stimulating growth by the use of artificial chemical fertilizers, causes his gardening to become unbalanced and unstable. In fact, his crops are obtained at the expense of the soil's capital, its fertility.

Every reader should do everything possible to keep up true soil fertility, and this means the keeping up of the supply of humus.

No book will satisfy everyone. Some people are so frightened of chemical fertilizers that they would not dare to put even a spoonful on the soil. Others like me believe that providing the soil is enriched with plenty of organic matter regularly, there is no harm in using a slight trace of some particular 'chemical' as a tonic. Please note this word 'tonic'. It may be necessary on occasions to give a fillip to a crop, or to make up for some known deficiency. Even then I would prefer to use an organic fertilizer like dried blood for adding nitrogen, bone meal for phosphates and wood ashes for potash.

In a recent report of Britain's National Vegetable Station these words occurred:

None of the combinations of inorganic fertilizer treatments give yields as high as from plots which have received Farmyard Manure. On average the latter gave a 75 per cent increase in yield compared with the best fertilizers' treatments. The beneficial effect of dung can be seen at a very early stage of growth. Farmyard Manure applications have increased the water-holding capacity of the soil.

In the case of gardeners, compost takes the place of old farmyard manure – and in fact is better. Every gardener can make good compost from the organic waste in his home and garden.

Read through the chapters carefully, and make up your mind what you will believe, and what you will do in consequence.

CHAPTER 2

The Soil: What It Is and How We Got It

Far too many people call soil 'dirt'. There is nothing dirty about the wonderful living material in which we grow our plants. This earth is a layer of disintegrated rock, which is fitted by nature under the proper conditions of climate to support plant growth. The type of soil differs in accordance with the rock from which it came, while the depth of the soil often depends on the contour of the surface and on the rainfall. Mountains which have a heavy rainfall undergo a certain amount of weathering and washing away of soil. The particles are then carried down the valley below, where, in consequence, the soil will usually have a much greater depth.

Soil does not only contain this disintegrated rock or minerals, but contains in humus organic matter in various stages of decay. Of course it also carries the millions of living organisms which carry out the many operations described in Chapter 3. In soil will be found also certain gases and moisture; and where there is any depth of soil, as there is on cultivated land, the top 180 or 200 mm (7 or 8 in) may be dark in colour and the soil below lighter. This darkness is largely due to oxidation and the organic matter present.

The minerals which interest the gardener are quartz and granite, which being practically insoluble in water are little affected by the weather. Very often the 'cementing material' of such rocks, ie the clay or felspar, do disintegrate by weathering, and fragments of quartz may therefore become detached. Felspar appears hard, but can easily be decomposed by weather.

Mica is a common mineral in soils, and can be disintegrated by the atmosphere. Quartz may be said to be of no value as a plant food, but felspar does help in providing potassium, lime and a certain amount of iron. Calcium carbonate, as found in such soils as limestone or chalk – or even marble for that matter – is another common mineral. It can be of value both as a plant food and as a releaser of other foods. Too much lime in the soil can, however, have a harmful effect on certain plants, producing that yellowing condition of the leaves, known as lime-induced chlorosis. This is caused by the lime preventing absorption of iron through the roots of the plant.

When I have been lecturing on this subject, the question of where the rocks came from has invariably been raised. The rocks were, of course, the original crust of the earth, but they have been through a succession of changes. The general classification of rocks is usually: (1) igneous, those produced by volcanic eruption; (2) sedimentary, those that have been formed under water; and (3) metamorphic, which have been produced by high temperatures and pressures, sometimes changing their primary characteristics. Two typical examples of metamorphic rocks are marble, which has been converted from limestone, and slate, which has been converted from the shales or clays. The normal clays, sandstones, and limestones, are soils in what is called the sedimentary group.

All kinds of agents have been the means of building up or breaking down the substance we normally call 'soil'. Wind, for instance, has transported matter from a distance. Frosts and thaws have broken up the rocks, and have had a similar effect on very heavy clays. Rivers have carried down silt, sometimes overflowing and depositing the silt on either side. Sometimes the rivers have altered their courses, and they have left silty river beds behind them. Glaciers in travelling down a mountain have ground down rocks to the finest powder. Gales and winds have caused further physical damage and rock disintegration.

Animals, too, have played their part, particularly the burrowing types. Foxes, badgers, rabbits, moles, and even voles, aerate and fertilize the soil. Earthworms have done a great work, not only by carrying soil to the surface, but also by drawing dead leaves into the holes and so increasing the humus content. Chapter 4 shows how worms play their part in improving the plant-food content of the soil.

Ants may be regarded as a nuisance in the garden, but in many countries peoples have found 'ant-heap' soil ideal for seedbeds, especially when mixed with a little silver sand. Many other creatures have played their part in helping to produce soil. Wireworms and cutworms, for example, are considered pests because they may ruin crops, but they do play some good part in the soil itself.

The vegetation itself does its work. The roots penetrate into the ground and aerate it, for when they die they leave a tiny channel down which the air and moisture can go. Plant roots also help to break up the rocks, both mechanically and chemically, while bacteria in the soil work on the organic matter and even on the mineral matter produced by the rocks.

THE FIVE MAIN SOIL TYPES

There are five main kinds of soil. These are (1) clay; (2) sand; (3) loam; (4) limey or calcareous; and (5) peaty or moss.

(1) *Clay* This is a smooth, silky soil to touch, which can be rubbed through the fingers without feeling gritty at all. Even when well drained it is apt to be wet. It is difficult to cultivate during rainy periods and in the winter months. If it is worked when wet it has the unhappy habit of setting like cement, and then is very difficult to work and break down afterwards. It is a good thing, if digging is done, to carry out work on a clay soil in the autumn and to leave it rough, because then the winter frosts and cold winds will pulverize the exposed lumps of soil, and so make them more easily workable in the spring.

Heavy clay soils are recognized as being late, because they are cold and take a long time to warm up in the spring. Therefore it is impossible to work them as early in the spring* as it is in the case of sandy soil. Clay soils, on the other hand, are richer in plant food than sands, and in a dry summer season are valuable because of their water-retaining properties. Clays should be given regular dressings of lime, because this helps to 'open them up' and make them more workable.

(2) *Sand*　A sandy soil is, of course, light and dry. It is easy to cultivate, and can be forked or hoed at any time of the year. It is called a warm soil, because it accepts the sun's rays much more quickly in the spring, and so can be got into production earlier than a clay soil. Sandy soils are, however, poor in plant foods, and especially potash, and they do not retain moisture easily. Contrary to popular belief, they can be very acid, and so need regular dressings of lime. They are invariably short of organic matter, and far heavier quantities of well-rotted composted dung or composted vegetable refuse have to be added each season than in the case of clay soils, if the humus content is to be kept up. In this way such soils can be encouraged to hold the moisture better.

(3) *Loam*　A loamy soil is the ideal blend of clay and sand. It has the advantages of the two types and none of their disadvantages. The presence of sand allows the water to percolate through quickly, and the presence of the clay helps to keep the soil moist. A loamy soil is sufficiently warm, and is not so lacking in plant food as a sand. Loams differ of course in accordance with the proportion of sand that is present. Like others soils, they may need dressings of lime, and they will certainly appreciate regular applications of fully rotted compost. The gardener is indeed lucky who has a loam to cope with.

(4) *Limey soil*　A chalky or limey soil is usually rather

* At Arkley Manor where no digging has been done in the last sixteen years it is possible to get on to the soil any time in the winter or spring because large quantities of compost have been applied each October–November. The soil concerned is a heavy London clay.

shallow. It is lacking in humus and unfortunately in plant foods. Such a soil will be sticky and unpleasant to work when wet; it is equally disappointing in a dry season, because it suffers so soon from lack of water. Plants growing in such soils often suffer from lime-induced chlorosis, reflected in yellow leaves and stunted growth. Chalky soils have one advantage, and that is that they are not acid, and so it is seldom, if ever, necessary to lime them. Much can be done with them, if vast quantities of organic matter are applied to the surface of the soil each season in October or November.

(5) *Peaty soil* In some parts of England, peaty soils are described as moss lands. They are often waterlogged, and so need careful draining, and as they are usually very acid, or sour, lime has to be added. Their one great advantage is that there is plenty of organic matter present, and so little compost needs to be added. The brown peat lands are much easier to work and crop than the black heavy bog-like peats.

It has been said that the top 225 mm (9 in) of good soil in a temperate climate contains enough plant foods for over a hundred crops. Unfortunately, only a small fraction of the food is in the condition in which plant roots can take it in. The fact, however, that there is a plot at the Research Station at Rothampstead, Hertfordshire, which has been cropped with wheat since 1844, without adding any fertilizers, shows what can be done. Unfortunately, the yields at present-day values are very low indeed – 300 kg (6 cwt or 6.7 US cwt) as against 1000 kg (20 cwt or 22.4 US cwt) – but it does show what nature will do even though the same crop has been taken from this land for 130 years! Whether of course the wheat produced has the same food value or not is not stated. This is an important point to remember.

THE PROFILE OF THE SOIL

If a pit is dug and it has one clear-cut vertical side, it is possible to see what is called the soil profile. This is the name given to the obvious layers of the soil which the geologist calls 'horizons'. These varied layers show the gardener the state of balance between the percolation process known as leaching and the amount of organic matter or compost given as replenishment. Thus it is possible to see what the worms have done and where the beneficial bacteria and the helpful fungi have been working.

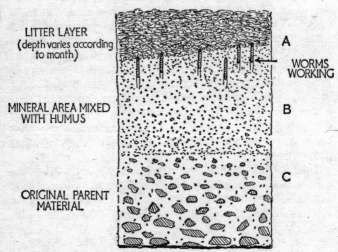

LITTER LAYER
(depth varies according to month)   A

WORMS WORKING

MINERAL AREA MIXED WITH HUMUS   B

ORIGINAL PARENT MATERIAL   C

THE SOIL PROFILE SHOWING SOIL ZONES

There are, you will discover, three main zones. Zone 1 usually contains the debris – the organic matter of the litter layer. In the wood the litter layer is comprised of the leaves and other undecomposed material of the forest floor. In an ordinary garden it is the compost layer of the 'no-digging' gardener. The litter layer is broken down by the activity of soil creatures, like mites, bacteria and fungi, and is incorporated into the microbiological layer (Zone 2) by the earthworms. In the microbiological layer the bacteria and fungi transform the material derived from the litter

22

layer into 'humic compounds'. These compounds are washed down into the macromolecular layer (Zone 3). The freer the drainage the more the leaching. Plant foods like potash, lime, magnesium and nitrates are passed down together with organic matter like humic acid, and this leaching can bring about acidity. This is one of the reasons why it is so important to give a top dressing of compost, because it minimizes the leaching. Where no compost is given, the soil surface is exposed, leaching becomes serious, acid conditions occur and the soil structure is impaired. If the gardener depends on giving lime and lime only to the soil, mineral deficiency occurs. Liming therefore is never a substitute for correct organic gardening. If lime has to be used because of acidity, then organic manures and composts must be used to maintain the soil in the right condition.

At The Good Gardeners' Association Research Station it has been discovered that liming has been unnecessary after the first two or three years. The vegetable waste takes up the lime and in the recycling process of composting the lime returns to the soil.

In every garden there is always an interplay of these main factors: (1) the parent material or organic soil which is available; (2) the climate in that season or year; (3) the organic life in the soil (see Chapter 3).

THE RHIZOSPHERE AND THE MICRO-ORGANISMS

The macromolecular layer is free of bacteria and fungi, and it accommodates most of the plant roots, which like the accumulation of humic compounds there. The striking resistance to disease common to plants grown in an undisturbed soil is due largely to the low numbers of bacteria and fungi in the macromolecular layer.

In this layer are found the greatest concentration of micro-organisms in the neighbourhood of the plant roots. The micro-organisms which are closely associated with the plant root are known as rhizosphere micro-organisms. By the word rhizosphere we mean the area of contact between plant root and the soil. Rhizosphere oganisms live on secretions from the plant roots – mainly sugars and amino acids. These result from photosynthesis and the plant's metabolism. In return, the rhizosphere organisms provide the plant with substances for its use by process of synthesis and decomposition. One example is the provision of nitrates by the decomposition of organic matter around the root.

Since the components of root secretions differ from plant to plant, the types of rhizosphere micro-organisms which feed on those secretions will also differ from plant to plant. Presumably each plant, through its root secretions, selects those micro-organisms from the general soil population most suited to supply its needs.

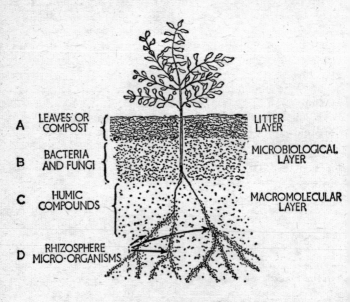

A LEAVES OR COMPOST — LITTER LAYER

B BACTERIA AND FUNGI — MICROBIOLOGICAL LAYER

C HUMIC COMPOUNDS — MACROMOLECULAR LAYER

D RHIZOSPHERE MICRO-ORGANISMS

ANOTHER DIAGRAM SHOWING THE
SOIL PROFILE WITH A GROWING PLANT

A fertile soil is one that produces large quantities of high-quality crops. If we make use of the idea of the rhizosphere, we reach a definition of soil fertility which provides the basis for the measurement of fertility. We shall say 'A fertile soil is one that responds to the presence of a plant root with a vigorous and beneficial rhizosphere population.'

CHAPTER 3

The Living Agents of the Soil

It is difficult to know how best to impress on beginners the importance of encouraging the living agents of the soil. One can say, quite truthfully, that in a normal handful of fertile soil there can easily be 69 million living organisms of one kind or another, and that there should be 55 tonnes of bacteria and fungi in the fertile top 150 mm of 1 hectare of good soil (equivalent to 28 million living organisms or 50 000 lb of bacteria and fungi in the top 6 in of 1 acre). The living agents may be infinitesimally tiny, but because they are there by the millions, they bulk up and have very definite weight.

Animals like foxes and stoats (and, in the past, wolves) not only help the soil by burrowing and scratching, but also by their predatory habits. Bones of other animals they have attacked decompose to produce plant foods, and their excreta, made more valuable by the eating of such flesh, eventually contributes too to the richness of the soil.

This chapter concentrates on what some people call the microbes – the organisms in the soil which are almost unknown, because unseen. Their work is mainly concerned with death, and with decay. It is tremendously important work, for decomposition leads to recomposition, and the provision of fresh material on which the plants may live. Some microbes can fix the nitrogen from the air, or can release it from decaying organic matter, so that it can then be converted into forms suitable for absorption by the roots of plants.

DIVIDING THE MICROBES

It is convenient, for the purpose of this book, to divide into six main groups the microbes or lowly organisms that have

26

to do with the improvement of the soil and the growing of plants. These main divisions may not be super-scientific, but they are suitable for such a book as this.

(a) *Yeasts, moulds and fungi* It is possible to have about one million fungi in 1gm of soil. Fungi can be said to be plants, but they are plants which have no green colouring matter or chlorophyll which is present in the foliage of plants which live above ground. Most fungi are parasitic, that is to say they live on other plants, for instance the mildew on cucumbers, the rust on chrysanthemums and the black spot on roses. Those that do not live on other plants have to find their food from the remains of recently dead plants.

The parasitic fungi on the whole are disease-producing, like those we have already mentioned. The healthy fungi as a class are saprophytic, that is to say they live on the dead and decaying matter in the soil, or sometimes above the soil.

The bluish growth of mould found on an orange may not look pleasing or helpful, but actually the fungus is trying to convert the locally damaged material of this fruit and so get it into circulation for the use of what may be called higher plants. A mould can never grow on an orange, unless the fruit has been opened up by the peck of a bird, or one side has been 'killed' by a blow. It is when this happens that the spores which are always present in the atmosphere find the right conditions in which to grow and are able to settle down on the affected part and do their work.

Fungi on the whole propagate themselves by sending out thin thread-like substances into the material on which they are living. These look like minute spreading roots which are known as mycelia (the singular being mycelium). Mycelia are not really roots, they are part of the fungus. Mushrooms are a typical example. The mycelia grow in the compost provided, and from these mycelia grows the edible

27

fungi or mushroom. This in its turn bears the spores (or seeds).

Spores which contain one cell (or nucleus) may be dispersed in the air and though they cannot be seen by the naked eye they may drift in the wind from place to place. Sometimes spores cover themselves with thick coats. These are then known as resting spores. They can wait for long periods – possibly years – in a resting place, until the right type of food is supplied, or until the right kind of conditions develop for the fungus to start into growth again.

The moment a fungus gets established on a plant, it not only feeds on the material but sends up tiny aerial stems on which the spores are borne. It is in this way that each particular fungus distributes itself over the land. Fungi can therefore work on dead material in the soil, and they can work equally well on the leaves of plants just above the soil, or on the stems of plants at soil level. There are, however, fungi like *Verticillium wilt* and *Fusarium wilt*, which will attack live roots of plants, and kill them.

(b) *Mycorrhiza* This name 'mycorrhiza' comes from 'myco' meaning 'fungal', and 'rhiza' meaning 'root'. Modern research shows that there are many plants which benefit from the presence of certain soil fungi in the soil. Plant roots are invaded by the mycelia of the mycorrhiza, and this lives partly inside the roots and partly in the soil. The mycorrhiza gets its carbohydrates from the root, and in return it gives the plant nitrogen, certain other nutrients and moisture. Later on, the plant roots will absorb the mycelium into their own tissues.

Dr N. C. Raynor, in her work on the Wareham soil, was able to show that fertility is bound up with the stimulation of mycelium activity, and that the only way of ensuring such activity is by the addition of properly composted vegetable waste to the soil. It may therefore be said that the mycorrhiza which benefit the roots of plants, actually thrive in the humus and not in the soil where there is none present. There is evidence to show that the *mycorrhiza are dis-*

28

couraged, and may even be inhibited, by many of the chemical fertilizers in use today.

If you raise the nitrogen content of soil by the use of, say, sulphate of ammonia or nitrate of soda you will certainly reduce and most likely prevent the fixation work of the mycorrhiza, as well, incidentally, as the work of the bacteria which causes those beneficial root nodules to be produced on members of the Leguminosae family (ie the peas and beans, clovers, lupins, etc).

Mycorrhizas co-operate with at least 80 per cent of our flowering and cabbage family, and the potatoes and tomatoes. It is not the actual plant nutrients in a compost that makes the difference. No cure can be achieved by the addition of inorganic fertilizers, and in fact the continued use of such fertilizers (Dr Raynor discovered this in her experiments) increased the difficulties of growing the plants rather than remedied them.

Properly rotted compost evidently encourages the mycorrhiza association in plants, whether in the form of a sheaf of mycelium on the tips of young rootlets which can be seen, and whose infection is merely inter-cellular, or in the inter-cellular form where finally the mycelium undergoes complete digestion within the cell walls. It is tremendously important to bear in mind these mycorrhizal relationships when considering the feeding of the soil.

The late Sir Albert Howard stated very clearly that

The long-continued use of artificials is followed by disease and the production of indifferent fruit. The strawberry is a mycorrhiza former ... and those who grow strawberries can very easily compare the effect of humus and chemical fertilizers on the health of the plants and on the quality of the fruit ... The mycorrhiza appear to be the machinery provided by nature for the fungi living on the humus in the soil to transmit direct to the active areas of roots, the contents of their own cells ... If, as seems almost certain, freshly prepared humus does contain growth-promoting substances (roughly correspond-

29

ing to vitamins in food), it would be necessary to get these into the plant undamaged and with the least possible delay. The mycorrhizal association in the roots, by which a rapid and protective passage for such substances is provided, seems to be one of nature's ways of helping the plant to resist disease.

GROUP 2

Bacteria Bacteria are microscopical, and because they are transparent it is generally necessary to stain them in order to examine them properly. Most of them are round, but some are rod-shaped, similar to the fungi, but there is never more than one spore, ie one cell. Some bacteria have phlagella, that is to say little tails, by means of which they can swim about in moisture. Bacteria may form themselves into chains. The rod-types increase quickly by division, a single cell breaking into two, these cells in their turn into two more each and so on. Other bacteria divide themselves into fairly equal quarters, and in this case reproduction is not only simple but extremely rapid. From one bacteria it is possible to produce 16 million more in twenty-four hours.

The bacterium known as *Bacillus radicicola*, for instance, not only lives, but multiplies on the nodules formed on the roots of leguminous plants. It receives free quantities of carbohydrates, and gives nitrogen gladly. Some of this nitrogen is used by the plant, while the rest is left in the soil for the next plant. The nodules on the American Sweet Clover, for instance, when grown in soil, can produce the equivalent effect on 5 tonnes of ammonium sulphate per hectare (2 tons or 2.24 US tons per acre)! The work done by these bacteria, therefore, ensures the addition of nitrogen in a natural manner, and makes it unnecessary and in fact *unwise* to apply chemical fertilizers.

The group of bacteria usually known as azotobacter do not break down, but build up. They get to work on the gases and other compounds that might easily be lost in the soil and they build them up into such a condition that they are held in the soil and are used eventually by the plants. About 63 tonnes of nitrogen may be fixed in the top 150 mm layer of 1 hectare of soil by these azotobacter (equivalent to 25 tons or 28 US tons in the top 6 in of 1 acre). The azotobacter require air and food in the form of decaying organic matter. That is one of the reasons why the gardener should give compost each season. One of the azotobacters is the *Bacillus radicicola* already described. All azotobacter play a great part in keeping up the fertility of land. They help to improve the physical condition of soil, because of the jelly-like celloids they produce, which cause the soil particles to stick together in the right way.

Protozoa The protozoan is about twice the size of what may be called a normal microbe, and many times the size of a bacterium. It can be found in all kinds of shapes, and what is usually seen under the microscope is a little speck of jelly. Most protozoa are not helpful to plant life. Many of them appear to be green in colour, because they are so often associated with the microscopic green plants known as algae. These algae probably provide the sugars which the protozoa need to live on. Some experts consider them to be animals by night and plants by day!

Algae Algae can be said to be plants without any true roots, and without distinctive leaves or stems. Some are very large, like the seaweeds, and others are microscopic, like the algae that live with the protozoa. What look like roots or seaweed are merely anchorages and they do not play any part in passing food up to the plants themselves. The chlorophyll (the green colouring matter in leaves) which the algae contain does however enable them to trap the energy of the sunlight. Algae are bathed in moisture all the time – seaweed and fresh-water algae live in ponds, and even the microscopic algae in dry soil are covered with a water film.

The cell of a baby alga divides up apparently spontaneously into two equal halves. The bigger algae are very useful indeed. Iodine, for instance, is extracted from seaweed, while if seaweed is ground to powder it makes a good liquid manure. Apart from the life of the algae with the protozoa (see above) it is not fully known what function they have in the soil.

(see above)

GROUP 5

Lichens Despite popular belief, the lichen is not a true parasite. It will be found living on the branch of a tree or a rock, but it always seems to prefer a moist place. It would seem to be a kind of big brother to the algae and the fungus. Since such a partnership produces a plant which can be easily seen, it could be claimed that a lichen is not a true microbe!

GROUP 6

Actinomyces These have been left until last because they are very difficult to describe. They are almost a cross between a bacterium and a fungus. It is said that it is the actinomyces which give that earthy smell to soil. It is probably the moisture on the spores of the actinomyces which causes soil scent to rise after a rainfall in the summer.

GENERAL REMARKS

Some readers may find it a little confusing to read about the various microbes and the jobs they have to do in the soil. There is nothing really to worry about, however, for if properly rotted compost is applied regularly the gardener's day-to-day problems are solved. The bacteria carry out their functions without any instructions from the gardener and without any pay! Some attack the cellulose in plants, while others actually use the substances like carbonic acid for their energy.

Some bacteria can be called the scavengers of the soil, while others are entirely saprophytes. These latter concentrate on the carbohydrates and proteins. There are, too, bacteria and fungi that work on the nitrogenous types of matter, breaking them down into simple substances of an ammonia type, upon which the roots of plants can easily feed.

Once one set of bacteria has finished its work, another set will then attack the material. Sometimes the first set of microbes is absorbed by the second. The bacteria must, as a group, set about preparing food for the generations of plants that are to come. Without their help, life would cease.

The soil therefore is not only a place where the roots of plants can anchor; is not only a vast storehouse of plant food; but is a large manufacturing centre where millions of living organisms help the gardener. Dig in the compost or any other organic matter or apply it on the surface of the ground and immediately the countless organisms get to work preparing the necessary food for the plants that are to come. No gardener needs to worry about adding bacteria or microbes to the soil. The important thing is to see that sufficient organic matter is added to the soil each year so that plenty of humus is assured.

Humus is the by-product of bacterial activity, and microbes always work at their best when they have plenty of material to work on. As humus is so important and as microbes can and will do so much for the gardener, the least we can do is to see to it that organic matter is added in abundance year by year.

It is often asked what happens when soil sterilization takes place. Providing the temperature does not exceed about 99 °C (210 °F) in the case of baking, and the normal temperature of steam in the case of steaming, the injurious fungi and protozoa and any non-spore-producing bacteria are destroyed while the ammonia-producing bacteria which are spore producers are not affected and are then able to multiply in a phenomenal manner and carry out unrestricted operations.

CHAPTER 4

Beneficial Worms and Their Function

No one can be concerned with any form of soil culture without being interested in the Lumbricidae family, or earthworm group. The earthworms are perhaps the most easily observed group of soil organisms, but during the course of the years there have been many conflicting ideas as to the part they play in the ecology of the soil. Charles Darwin started the serious study of the function of the worm as long ago as 1830, and in the year 1837 he read a paper on the subject before the Geological Society in London.

His book *The Formation of Vegetable Mould* (1881) had a very big sale at the time. One can only regret that his painstaking investigations failed to influence teaching in agricultural colleges. It was a book, of course, which was far ahead of its time. In 1945 Sir Albert Howard reintroduced the book, mentioning in his new preface Dr Oliver, who in 1937 published in the USA *Our Friend, the Earthworm*.

This American expert has been able to restore fertility to barren land in hundreds of farms to which he has been called as an adviser. He solves the farmers' problems by the distribution of worm egg capsules. These restock the derelict land, which because of bad treatment is completely devoid of these valuable creatures.

The earthworm is naked and blind, and has no teeth or claws or any weapons of defence or offence; it has no mind to be afraid, and no feet to run away, and yet Charles Darwin wrote: 'It may be doubted whether there are any other animals which have played so important a part in the history of the world, as have these lowly organized creatures.'

It is the earthworm which continually renews and maintains the valuable film of top soil. All the waste products of life, the dead vegetation, the manure and dead animal residues, are the chief source of earthworm food. It has been said that animal life, in all its forms, from man down to microbe, is the great transformer of vegetable matter into food for the earthworm.

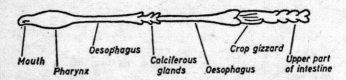

Diagram of the alimentary canal of the earthworm.

The number and distribution of earthworms is of course influenced by the many factors of soil environment. The writer has discovered tremendous numerical differences between adjacent gardens. No one can say that the distribution of earthworm populations is constant – the numbers will change in accordance with the supply of organic matter; the acidity of the soil, its moisture content and, to a certain extent, its temperature and texture. No one who is cultivating the soil can dare rest on his laurels and say, 'My earthworm population is high, and I need do nothing more about it.' The gardener must continually care for his soil, creating an environment suitable for the earthworm.

As there can be wide variation in the distribution of worms, from garden to garden, so there can be changes in what may be called the level of activity of the worms. These creatures are perhaps most active during the spring and autumn, when the conditions suit them best. They don't much like the dryness of the summer, nor do they like the low temperatures of the winter. Under such conditions they may make themselves little 'nests' in the earth, where they will curl up, and remain inactive until the favourable conditions return again. The gardener can encourage their activities by adding organic matter, which will form humus and so help to hold moisture, and of course by using some form of overhead irrigation.

Just as the level of activity may alter, so one usually finds two types of earthworms in a garden – one which seems to devote itself to the surface of the soil, and the other which burrows deeply and provides the channels down which the air and water can go. Some, as it were, do the deep forking, while others carry out the shallow forking or raking. The earthworm most met with by gardeners is the red *Lumbricus teyristers* which is often as long as 300 mm (12 in) and can have a diameter of 12.5 mm (0.5 in). This earthworm is a grand humus-producer and it deposits its excreta on the surface of the soil as a rule. It is, however, the deep burrowing, greyish or greyish-blue species that produce the majority of the casts, and the most minerals, and therefore add the greatest quantity of plant foods to the soil in the right condition for plants to make use of it.

The actual numbers of earthworms per acre, as we have said, may vary considerably. Darwin found concentrations varying from 62 000 to 131 000 per hectare (25 000 to 53 000 per acre). Dr F. E. Bear, of Ohio University, found worms present on the State University farm in numbers averaging 2 500 000 per hectare (1 000 000 per acre). Dr W. L. Bowers of the Oregon Agricultural Research Station writes of worms in the soil there varying from 62 000 to 3 700 000 per hectare (25 000 to 1 500 000 per acre). Dr W. J. Guild of the University of Edinburgh says that more recent research work has shown that field populations of between 1 250 000 and 5 000 000 per hectare (500 000 and 2 000 000 per acre) are quite common. Professor I. L. Heiberg of the New York College of Forestry stated in a broadcast talk that there could be more than 6 200 000 worms per hectare (2 500 000 per acre), weighing about 1600 kg per hectare (1400 lb per acre) and this meant that the farmer could have a greater weight of earthworms on his land than the weight of all his domestic animals put together. The late Mr Basil S. Furneaux, the famous British soil surveyor, has put this fact another way, saying that in the case of a well-manured pasture the weight of the worms in the soil will equal the weight of the bullocks grazing on the grass above.

In New Zealand, the Department of Agriculture has revealed that preliminary counts of earthworm population in fields of the well-managed country sheep farms give numbers ranging from 3 700 000 to 8 600 000 per hectare (1 500 000 to 3 500 000 per acre). The grey British worm, *Allolobophora caliginosa*, is said, incidentally, to be ten times more effective in New Zealand than the native worm.

From the weight of casting thrown up by earthworms in the Valley of the White Nile, agricultural experts in the Sudan have discovered that during the six months of the active growing season the weight of the worm casts work out at 120 tonnes (119.79 tons or approx 134.4 US tons) per acre. This is an extraordinary figure, because it means in fact that the probable output of worm castings for the year is more than 200 tonnes (200 tons or 224 US tons). There is no doubt that the great fertility of the Valley of the White Nile is partly due to the work of earthworms. The weight of the worm casts deposited there each year is ten times greater than the weight normally deposited in Great Britain.

Dr C. L. Curtis of the Connecticut Experimental Station has proved that worm casts contain five times more nitrogen, seven times more available phosphates, eleven times more potash, and 40 per cent more humus than is found in the top 150 mm (6 in) of soil in which the worms may be living. How the worm by eating the soil can make it far more fertile in this extraordinary manner, no one actually knows. The organic matter and soil is neutralized in the alimentary canal by constant additions of carbonate of lime from three pairs of calciferous glands near the gizzard, and it is here that the matter is finally ground prior to digestion.

Practically all forms of organic matter are ingested by the worms, and this with the soil is passed through the gut until ejected as a casting usually on the surface, but sometimes below the surface, either as a lining to the burrows or just as a casting similar to those above ground.

If the field is rich in humus, the weight of worm casts deposited on the surface in the UK may easily exceed 4.5 kg (10 lb) per year per worm and it therefore does not take a mathematician long to work out the percentage improvement in the surface soil brought about.

It is suggested that there could be 20 000 000 earthworms in a single hectare (8 000 000 in an acre), and that some of their burrows might penetrate to a depth of 1.8 m (6 ft). One can imagine what effective aeration these creatures would achieve under these circumstances. The wide systems of small and large burrows which they develop through the soil help the excess water to get away as well as giving easy passage to plant roots and facilitating the important interchange of gases between the atmosphere and the soil itself.

A gardener should always examine the worm population during his digging and forking, for he can learn much during the process. If the worms in the soil are red, plump, glistening, active and about the thickness of a man's little finger, then all is well. There should be at least one good worm per spadeful. If on the other hand the gardener finds few worms, or he discovers that all he sees are pale in colour, sluggish and curled up in balls, then undoubtedly he is in for trouble, unless he adds ample quantities of organic matter.

If you could take a cross-section of your soil, you would see that all the parent holes have small caves or galleries excavated from their shafts radially, leading away about 5 to 12.5 mm (0.38 to 0.5 in), and sealed with castings darker than the inside of the shaft. At the end of some of the galleries you might see a small red bundle about pea size – a tiny moist knotted wormling – or an unhatched capsule or worm egg.

In the USA, millions of earthworms are raised in earthworm farms for selling to farmers and gardeners. Even amateurs breed worms under simple conditions in Britain. Boxes 355 mm wide, 445 mm long and 150 mm deep (14 in ×17.5 in×6 in) deep are filled with composted vegetable waste, and egg capsules are then 'sown' in the compost an inch deep. The boxes are stacked in tiers, and the temperature in the room in which they are stored should be between 16 and 21 °C (60 and 70 °F). The capsules then hatch out.

The young worm develops quickly, and may reach the reproductive stage in between sixty and ninety days. It will then begin to produce capsules on its own. This of course does not indicate that the worm is fully grown. During the whole of this time the compost must be kept moist, but not soggy, and the earthworms will then multiply rapidly. In another three months the creature should grow to its full size, when it should be about 100 mm (4 in) long. It will remain that size for many years, and at least one expert has been able to show that a healthy worm may live fifteen years and more. The worm breathes through its skin and it ejects the waste matter through its skin also.

No worm has to select or chew its food, all it has to do is to swallow anything that is small enough to enter its mouth. Furthermore, because there is both a male and female element within one body, it is quite possible to establish a colony of worms from either a single egg capsule, or just from one fertile earthworm. When The Horticultural Training College moved from Hextable near Swanley, Kent, to Thaxted, Essex, the soil in the new gardens was almost deficient of worms, and egg capsules were brought in for distribution in certain centres. The worms multiplied rapidly and within two years the complement in the college gardens was very high.

Another very useful function of the worm is to pull into the soil any organic matter deposited on the surface of the ground. It will drag down dead leaves and other parts of plants into its burrow, partly as food, and partly it would seem when it needs to plug up the borrow. Many a gardener has wondered why his shallots have moved after planting. The reason is that the worm, finding some loose skin surrounding a bulb, has caught hold of it to pull it into the ground and moved the shallot. The writer has heard allotment holders claim quite seriously that shallots jump in the night!

Once a worm has dragged a leaf or a piece of grass into

its burrow, it tears it to shreds, and partially digests it. It then saturates the remainder with intestinal and urinal excretions, in order to mingle it with the earth. Worms will alter the physical properties of soils. On light land, for instance, they have the effect of promoting particle aggregation and water-retaining capacity, while on the heavier clays they open up the soil and make it easier to work.

Darwin ends his excellent book with the following paragraph:

> The whole of the superficial mould over any expanse has passed and will pass again every few years, through the bodies of worms. The plough is one of the most ancient and valuable of man's inventions; but long before it existed the land was regularly ploughed, and still continues to be ploughed by earthworms.

The author would like to add to this sentence the five words, 'if given the right conditions'.

The use of chemical fertilizers on an extensive scale undoubtedly causes a reduction in the earthworm population. There was a well-known farm near the author's headquarters where chemical fertilizers were used in a big way, and where as a result it is almost impossible to find a worm!

Much has been said in Chapter 3 about the value of fungal activity, and it has been shown empirically that any conditions which discourage earthworms also prevent the perfect development of the mycorrhiza which associate with plant roots. If it is true that chemical fertilizers (and notably sulphate of ammonia) discourage both worms and soil fungi, then this undoubtedly is an important reason why they should not be used. Dr C. S. Slater of the USA Department of Agriculture stated in 1954 that 'the use of sulphate of ammonia for three years in succession on certain plots resulted in the decimation of the earthworm population'. This was also seen between 1949 and 1950 on a farm in Thaxted.

Organic manures of all kinds on the other hand definitely favour the development of worms. Research in New Zealand, for instance, found that the use of manure caused an increase of 13 per cent in the worm population, and increased also the total weight of worms in the soil by almost one-third. It was also revealed that insoluble molybdenum in the soil can only be rendered available to the plants by being passed through the guts of worms.

Lime does help in encouraging the development of earthworms *and* their activity, but only in the case of acid soil when a *moderate* dressing may be given. After three or four years of compost gardening lime is never used because the calcium taken up by the plants is passed back into the soil by the compost each season. Basil Furneaux used to say, 'The handling of soil is an art and not a science – and in the arts there are no hard and fast rules, but sound underlying principles. Each soil has its own personality and peculiarities. There is no virtue in cultivation – it can do more harm than good.'

The reason that digging, ploughing, forking and harrowing are harmful is that they break down the correct structure of the soil. Most people believe that the soil is a solid homogeneous mass – it is not. It is a delicate and complex structure. It must be in such a condition that it will allow for a ready entry of air and water. Few crops penetrate deeper than 0.9 m (3 ft) though beet may go down to 3 m (10 ft). This doesn't mean, however, that anyone need cultivate to that depth! In fact, the less the farmer or gardener cultivates, the better chance there is that the roots will do the necessary work assigned to them. Remember they must be able to breathe in oxygen, and give off carbon dioxide. If we are to produce edible crops under as favourable conditions as possible, then we must reckon that we should regard the plant as a whole. What is seen above the ground is one-third of the whole; we must, therefore, always concentrate on that part which is out of sight.

The air which the roots breathe and the water they need can only find their way between the minute particles of soil with difficulty. They depend on the highly developed system of communications produced naturally by the worms, the roots of other plants and the living organisms. Cultivations easily destroy nature's carefully made 'highways and by-ways'. The Good Gardeners' Association acknowledges that it is the work of countless organisms that are the true cultivators of the soil – not man.

The bacteria which live in the intestines of the earth-worms produce gums which stick the particles of soil lightly together and produce the correct crumb structure.

The organisms are the true cultivators of the soil and the gardener must see to it that they are helped and not hindered by digging, ploughing, or even deep hoeing. The truly fertile soil exhibits a crumb structure. The particles of soil are loosely stuck together to form what are called crumbs – these are only lightly attached to their neighbours. The hosts of fungi inhabiting the soil weave fine strands among and about the crumbs and so help to prevent them from coming apart or sticking together too strongly. The more fibrous the roots are, the most lasting their effect.

It is the long straight burrows of the earthworm that provide the main channels for the air and water. They may easily penetrate to 1.8 m (6 ft) and 2 m (7 ft). Roots find it far easier to go down a worm-hole than to go down through the soil itself. All the gardener has to do is to supply the organic matter in the form of compost and this compost should, by its very make-up, contain lime.

The greatest mistake a grower can make is to break up the natural system of communications. If he does this, for instance when preparing a seed bed, the result will be that the stirred part will lie wet and unfit to touch long, long after the unstirred area is ready to be sown. I have often seen water lying on top of a sandy soil causing, after drying, a caked or panned surface. There are no crevices then for the air to enter and the roots will soon be literally gasping for breath. Use a rotary cultivator by all means but never deeper than, say, 50 mm (2 in); always leave the lower soil structure as it is. The shallower the cultivation of the soil the better.

Readers should therefore encourage worms in their garden, because they help to speed the breakdown of organic matter; they burrow and materially aid soil drainage, allowing air to penetrate more deeply; they turn over the soil and enrich it at no cost to the gardener; and they have a profound effect on the ecology of the soil.

CHAPTER 5

Nature's Wheel of Life

The name Adam, which God gave to the first man, means 'of the ground'. The name was not given thoughtlessly, for we read in Genesis, Chapter 2, 'The Lord God formed Man out of the humus of the ground, and breathed into his nostrils the breath of life, and man became a living soul.' Then later on in Genesis 3, verse 19, we read that God said, 'In the sweat of thy face thou shalt eat bread, until thou return to the ground, for out of it wast thou taken; for humus thou art, and unto humus thou shalt return.' The actual word used in the Authorized Version is the word 'dust', but man's body, when it rots down, does not form dust, it forms humus. Jeremiah makes this very clear in Chapter 9, verse 22, when he writes: 'Even the carcases of men shall fall for dung upon the open field.' Dung (organic matter) eventually becomes humus.

We gardeners have a duty to follow this divine plan, as well as a duty as stewards of the soil. We must do everything possible to conserve the resources of the earth, and increase its productivity, so that we may pass on its value and potentiality from generation to generation. There is afforestation to be done, for instance, soil erosion to prevent, and in the garden there is the passing back to the soil of the humus which the crop has taken out. Nature's plan is to build up the humus year after year, and this can only be done by organic matter. Unless that which has been taken out is replaced, soil sickness occurs.

The Chinese have always taken an interest in the fertility of their soil, and after 4000 years of intensive cultivation still support more human beings per acre than any other country in the world. Because they collect and use and return to the soil every possible kind of waste, vegetable, animal and human, the fertility of the soil has not

45

deteriorated, neither have the crops it produces. The use of artificial manures alone may seem to succeed for a number of years, but the humus is steadily depleted until little or nothing can be done.

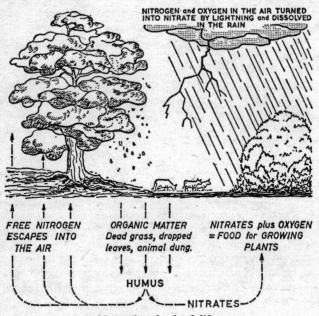

NITROGEN and OXYGEN IN THE AIR TURNED INTO NITRATE BY LIGHTNING and DISSOLVED IN THE RAIN

FREE NITROGEN ESCAPES INTO THE AIR

ORGANIC MATTER Dead grass, dropped leaves, animal dung.

NITRATES plus OXYGEN = FOOD for GROWING PLANTS

HUMUS

NITRATES

Nature's wheel of life.

One obvious example of Nature's plan is the tree. It produces a crop of leaves in the spring, which through the action of the sunlight manufacture starches. These, converted into sugars, are carried to the various parts of the tree as food and then in the autumn, when their function is over, the leaves fall to the ground. The worms pull these leaves into the soil and so return to the living earth the organic matter it so badly needs. Thus the circle is completed.

46

The same thing happens with the grass in the field. As it dies in the autumn and winter, it is pulled down into the ground. Maybe it is eaten by the cow or sheep, and in this case the circle is completed by the animal dung deposits which pass down into the soil to produce more humus. Death is Nature's way of providing for new life. As life gives way to death, death gives way in its turn to decay and decay is the means of new life again.

Not only do the leaves of trees and the older grass blades, etc, fall to the ground but also the dead birds, animals, flies, beetles and so on, all of which are acted on by the appropriate organisms and converted into humus, making it possible for fresh life to spring out of the soil. There must be that perfect balance between decay and growth, and it is when this takes place that the gardener can ensure that the fertility of his soil is perfectly maintained.

It must be remembered that every piece of work done in the garden or on the allotment helps to break down the humus, and every crop grown reduces the organic content of the ground. Thus constant care has to be exercised to ensure that humus is replaced. There can be no permanently perfect soil. There must be change, and the gardener should aim at having a vigorous, living, breathing soil, a soil from which the capital is never removed, and the capital is, of course, the humus content. When buying land or when renting a garden it isn't the mica particles nor the grains of sand that are the value. It is in the humus that the soil contains.

If land has been ruined year after year by constant chemical manuring and by the lack of application of organic matter, then obviously the soil will have decreased in value. The old adage says, 'Lime and lime without manure will make both farm and farmer poor', and the manure mentioned there is farmyard manure and not the artificial fertilizer.

HUMUS – WHAT IS IT?

The end product of the compost heap is humus. Its

definition is problematical. One can say that its value lies in its coagulating effect on the soil, and on its water-holding capacity; that it is a complex residue of partially oxidized animal and vegetable matter, together with substances synthesized by fungi and bacteria; that it is the key material of the wheel of life; that its effect on crops is nothing short of profound. One can be even more erudite, and say that humus is a composite entity possessing biological, chemical and physical properties, which make it distinct from all other natural organic bodies. Even then, humus has not been fully described, its character is so complex.

It is there to be a food for the other micro-organisms and other creatures. The organic matter provides that. It is formed from the work of worms, animals, insects, innumerable different kinds of bacteria, myriads of different types of live and dead organisms such as moulds and yeasts, and includes the residues of plants and animals. It is the relatively stable substance which is left behind after the bacteria and fungi have fed on the organic matter worked into the soil.

During the formation of humus all sorts of chemical changes take place, and the soil organisms work on the proteins and carbohydrates and break them down into simpler substances. Gases are formed in the soil during the process, and these are the cause of further work. Humus must be continually replenished, and then it will effectively maintain garden soils in a suitable physical condition.

The great thing to remember is that humus isn't 'dead' in the normal sense of the word. It is the transition stage between one form of life and another. Remember that there is an organic cycle and that there are constant changes and constant processes going on. No horticultural chemist can analyse humus and tell us what it is – wherever there is life there is change, and the author refuses to believe that the human body is just some chemicals and a quantity of water.

It has never been claimed by those who believe in the use of humus that compost will provide all the necessary nitrogen, phosphates and potassium needed by the plants to be grown, or that the chemist is wrong when he estimates the requirements of plants in the form of nitrogen, phosphates and potash. What the expert who advocates humus says is that the compost does feed the soil population; does help with the growth-promoting substances; does encourage the earthworms to do their job; does ensure the function of the mycorrhiza, and, in addition, does see to it that the soil organisms complete their own life cycle, providing plant foods by the decomposition of their dead bodies.

The chemist may argue that as all forms of nitrogen must be converted into nitrates before they can be used by the plant, to give nitrates in the form of a chemical like nitrate of soda merely short-circuits the business. He may claim that a dressing of such a chemical is far better than waiting for compost or dung to do its work. What he forgets, however, is that there is a very definite functional difference between the two. Take the baby's digestive ferments, for example. If the trypsin and the pepsin are transferred from the baby's inside to a test tube, they will digest a portion of steak and onions. Leave them, however, inside the baby and they can only digest satisfactorily the mother's milk. It is a question, surely, of natural processes every time, and not of chemical equations.

Incidentally, Lord Portsmouth invariably uses straw for thatching his cottages on his Hampshire estate in southern England. For a number of years now he has kept careful records, and although he has grown crops side by side on the same type of soil, he has discovered that thatch made with straw from wheat grown with plenty of organic matter to produce humus lasted twice as long as that grown with chemical fertilizers.

A number of British commercial market gardeners have had remarkable results with compost. Captain R. G. M. Wilson, of Surfleet, Lincolnshire, was much troubled by the uneven ripening of tomatoes, and tried the fertilizer treatments advocated by certain expert consultants without success. When, however, he adopted organic methods, this trouble disappeared and he has been able to grow crops of tomatoes in his greenhouses for thirteen or fourteen consecutive years without either sterilizing or changing the soil, and there has been no falling off in quality or in yield. Mr John Deacon, of Desford, Leicestershire, has planted tomatoes in undug soil in a big tomato-house for over ten years. He has given no artificial manures during this period, nor has he carried out any spraying. His method is to flood the soil thoroughly and then cover it with a 100 mm (4 in) layer of compost before planting out. He grows perfectly healthy plants as a result which are not only vigorous, but very heavy croppers.

It is very difficult to explain what humus is, but it does make all the difference to successful gardening. Eat a good cake and you may not realize that the eggs are present, but they have helped to make the cake what it is. Have plenty of humus present, and the soil is in good tilth. It has the 'paste' or 'jelly' present which prevents soil from being just dust. Humus is the organic colloid of the soil. When there are no organic colloids present the soil is useless. Humus is a substance made from material that was once alive. It can hold water; it can store plant foods; it can prevent valuable minerals from being washed away; it helps to keep the soil open; it helps to ensure the right aeration; and it gives the ideal insulation against heat and cold.

It is hoped that readers will do everything possible to ensure that their soil is enriched with humus year after year. We must speed on the making of humus, the conservation of humus, as well as our knowledge of humus, and thus see that our humus cycle is complete.

RAIN AND THE WHEEL OF LIFE

This chapter has already described the complete circle of Nature. The leaves falling down on the ground, the worms pulling the leaves into the soil – the living agents in the earth manufacturing, so to speak, all the necessary foods needed by the trees and the roots pumping these

50

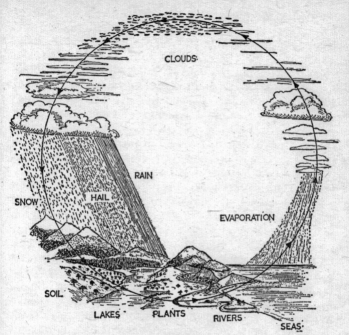

The water circle.

foods up into the tree again. Everything that has lived in fact can live again in another plant and so everything possible is composted and put back into the soil.

Another 'complete circle' is, of course, electricity. Turn the switch down and the circle is completed and the light comes on. The universe too is composed of circles – the earth goes round in a circle and the globe itself is a circle, is it not? But few people realize that the water we need is supplied through a natural cycle also.

The rain falls on the soil and gradually works down to the drains and ditches, and via underground channels, the rivers and seas. Meanwhile evaporation is taking place from the surface of the soil, from the surface of the water, and from the leaves of the plants – which, of course, are drawing water out of the soil. This evaporation eventually becomes condensation, and, by cooling through lifting higher and higher, the water vapour is collected in cloud-form until it falls as rain once more. Cold air cannot of course hold as much water vapour as warm air. Sometimes two air masses at different temperatures will meet forcing the warmer air over the colder. Large sheets of flattish cloud are then produced and rain falls for several hours. Alternatively moist air can be cooled by being blown over a tree-covered mountain or hill, causing low clouds and drizzle.

It must be remembered that a sandy soil will let the rain-water through quickly and easily but that heavy clay soil gives resistance to the downward penetration and in some cases the rain that cannot penetrate runs off to waste on the surface. It is important to open up a heavy soil by giving 25 or 50 mm (1 or 2 in) mulching with powdery, dark brown compost or medium-grade sedge peat. The worms will thus pull in the organic matter needed to build up the humus content of the soil 0.9 or 1 m (3 or 4 ft) deep or more – and thus the water will be able to penetrate. If enough water is not given then the roots come upwards to reach the surface moisture, and then, if there is little rain or in-efficient watering, the surface of the soil will dry up and the roots will suffer badly.

CHAPTER 6

Organic Manures and How To Use Them

In the peace and quiet of a forest, one may be looking at an old oak tree that has been growing there for some 500 years, and realize that the roots of the tree have not exhausted the land, in fact the leaves that have fallen from the branches every year have re-enriched the soil, in accordance with the cycle described in the previous chapter. On the other hand, when digging up the garden and disturbing the soil, and when carrying out continuous cropping and inter-cropping, there is always need for adding organic matter to the ground, and this, in a modern world, is often a problem.

Before the days of steam, petrol, oil and atomic power, transport from place to place was largely effected by means of the animal. Generally speaking, it was the horse. Those who owned large stables in the towns or cities were glad to give the manure to market growers and gardeners who came to fetch it. This was the one way they had of getting it cleared away.

A market gardener who took his fruits, flowers and vegetables into the town two or three times a week, invariably came back with a load of horse manure. It was this regular traffic backwards and forwards that made the production of first-class vegetables possible. The market-garden lands near towns therefore improved year after year. Incidentally there are still market gardeners who do this, and the royal stables at Buckingham Palace are regularly cleared by a grower from Milford, Surrey. Unfortunately, most of these perfected soils, rich in humus as the result of regular heavy dunging, have been built over, for until the last few years no one seemed to care about the conservation of good soil. Years ago the best salads and vegetables were grown at Camberwell and Mortlake in Surrey, but today the fine soil is now covered with houses. The author had a part to

53

play in an unsuccessful attempt to prevent the ruination of the most beautiful market-garden land at Baguley in Cheshire, when Manchester decided to use this area for a building scheme. Once again, the most wonderful earth, rich in humus, is covered with houses.

In the olden days farmers throughout the country tended towards what was called general farming. They kept cows for milk; they fattened bullocks in the yard; and they kept pigs. Thus there was always plenty of manure year after year to plough into the arable land. If farming is to succeed in the future, some such scheme will have to be compulsory, because at the moment hundreds of farmers are ruining land by burning straw, and by using chemical fertilizers. An alternative scheme would be to lay the land down periodically as a temporary pasture, so as to be able to plough the turf in shallowly at the end of the year.

It would be better, however, to legislate that all farms should have a herd of beef cattle. This would mean that the young beasts would have to be cheaper, and the finished beasts would have to receive a fair guaranteed price. There should always be plenty of straw about, so that the bullocks can be wintered in a yard, with straw litter in plenty, and thus the dung problems would be solved. When these are solved the fertility problems are solved also, and soil erosion is prevented.

FARMYARD MANURE OR DUNG

This consists of the excreta of the animals, both solid and liquid, plus the litter that is used for the animals, usually straw, and any waste food materials. Plenty of litter should always be used so that the liquid portion of the excreta may be soaked up; broken and bruised litter absorbs the urine much better than straight uncut straw. Naturally, the value of such manure depends on a number of conditions: the type of animal kept, the age, whether the animal is being worked or fattened, and so on. The better the animal is fed the better the dung. Probably a very high proportion of the total manurial value of farmyard manure is contained in the urine of the animals, which the litter soaks up. It is important to see that this soaked-up urine is retained, and a proportion of its value not lost, as so often happens in the heap.

As has already been explained, the litter enables the valuable liquid portion of the excreta to be retained, but it also provides the right type of material which can be converted into humus. It helps with the various fermentation changes that the excreta have to undergo, both because of its porousness, which helps to provide air, and because of the micro-organisms it supplies.

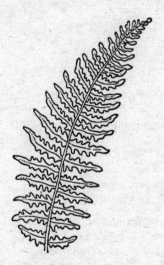

Bracken frond.

Straw is perhaps the best litter of all, especially when it has been well broken up. It is said that denitrifying organisms are found on the outside of the stem of straw. It is not a good thing, however, to cultivate strawy manure into the soil, for there may not be sufficient energy within the material to supply the millions of micro-organisms which must attack and break it down. Strawy manure, therefore, may cause nitrogen starvation for a while, the nitrogen being taken out of the earth, in order to provide the energy for the dispersal of the carbon in the straw.

Sedge peat is sometimes used as a litter, and possesses great absorptive powers for the liquids. Peat prevents the too rapid rotting of the organic matter of the excreta, and it is said that the manure produced is as a result richer, especially in nitrogen.

Dry bracken is often used in the north of England. It is not so absorbent as the other litters, but it does help to push up the potash content of manure.

Sawdust has good absorptive powers, but its very porosity leads sometimes to rapid fermentation to a harmful degree. This takes place particularly with horse rather than with cow manure. The sawdust from pine trees should be avoided if possible, because its decomposition is particularly slow, owing to the resins which it contains. On the other hand the powdered bark of pine trees has proved useful.

What has been said previously with regard to strawy manure is even more serious in the case of sawdust. Straw may have a carbon/nitrogen ratio of 30 to 1, whereas sawdust has this ratio at 250 to 1. This definite nitrogen robbing is therefore very much more serious. I have known it to last for nine months. Some gardening writers go as far as to say that it may last for years!

MAKING THE MANURE HEAP

The micro-organisms soon start to work on fresh manure in order to change its character. It is never advisable to dig in strawy manure because this may have a denitrifying effect on the soil – that is to say, the bacteria in the soil will have to work on the straw in order to break it down and thus less nitrogen will be available to plants. The manure should be kept in heaps until it is rotted, and the straw has shortened. The gardener often says keep the manure in a heap until it is 'sweet'.

While keeping, however, all precautions should be taken to prevent the loss of ammonia gas (for thus much ultimate nitrogen may be lost) as well as the loss of many plant foods by liquid drainage. The drainage water that comes away from a heap is very valuable, and should either be collected in a tank or should be collected and poured back on the heap again. The manure heap should never be kept in a loose porous condition, because then much of the ammonia will escape. It has been shown that in a carelessly kept heap of animal manure, air or rain has caused the reduction of its manurial value by half in eight months.

Make the heap in a sheltered place. If it can be kept in a covered bin so much the better. It should be packed tightly and the top of the heap should slope slightly to a point so that the water drains away easily. When it is well rotted and composted, it should be used at the rate of one good barrowload to 10 m² (12 sq yd). When well rotted, it can be lightly rotivated into the ground, with a rotary cultivator.

For ease of explanation all animal manures have been classed together as farmyard manure. It should be stated, however, that horse manure, or stable manure as it is sometimes called, is richer than cow manure, and it is particularly useful for making hotbeds, for producing mushroom beds and for the heavier type of soils. Horse manure loses far more in long keeping than cow manure. Cow and pig manure are very useful on the light sandy soils. Pig manure is intermediate in value between the cow and horse. Sheep manure, whenever available, is particularly rich in nitrogen.

OTHER ORGANIC OR NATURAL MANURES

Where bulky manures are needed (and all soils require them), it is possible to use one or two so-called substitutes. Remember that there is always the all-important compost, but a special chapter is devoted to this.

SHODDY

Shoddy is wool refuse and often used as a bulky manure for incorporation like dung. It should always be purchased on an analysis basis, for the nitrogen content may vary from 2 per cent to 15 per cent. The lower grades of shoddy are hardly worth having, except for their bulkiness. Shoddy acts like a sponge and helps to hold moisture. It is slow in action, but it sometimes contains weed seeds. It is usually necessary to apply lime as a top dressing after using shoddy. This helps in its decomposition.

SEAWEED

This is a cheap and valuable manure for those who live near the coast. It is comparable to farmyard manure, though slightly deficient in phosphates. It is quite free from weed seeds and is much better than farmyard manure in this respect. In addition it ferments more quickly. It is undoubtedly an ideal organic manure and should be used at a similar rate to farmyard manure or composted vegetable refuse. It should be incorporated at the rate of one good barrowload to 8 m² (10 sq yd). If bone meal is used at 100 to 135 g/m² (3 to 4 oz per sq yd) in addition, the deficiency of phosphates can easily be made up.

SEWAGE SLUDGE

The value of this product lies in the organic matter it contains for improving the physical condition of soil, rather than in its content of plant foods, for it usually has an analysis of about 2 to 3 per cent nitrogen, and 1 to 2 per cent phosphoric acid. It is used by some growers to mix with the vegetable waste on the compost heap and does add to the bulky material which can be dug in at the end of the rotting-down period.

HOPS

Spent hops are often used to make up a hop manure. This usually contains 2.5 per cent nitrogen, and normally no phosphates or potash, unless these have been added during the process of manufacture. A hop manure may be used as advised in the case of sedge peat, or may be dug shallowly in at the rate of one good bucketful per square metre (sq yd).

BIRD MANURES

When used as activators on straw and vegetable refuse, these form an excellent substitute for dung; but when used dry, they take the form of an organic fertilizer. Poultry manure has come into the market as the result of the large poultry farms which are being developed all over the country. Households have wondered how best to use this manure in consequence. It can, of course, be stored in old dustbins or barrels and be allowed to dry, when it is worth, as far as plant foods are concerned, about four times as much as the farmyard manure. If, on the other hand, it is kept with the straw or peat used on the dropping boards, then it is usually stored in a heap to rot down as ordinary dung, and so worked into the ground in the normal way. The dried powdery poultry manure is raked or hoed into the top 25 mm (1 in) or so of soil, as a top dressing in the spring, at the rate of 540 g/m² (1 lb per sq yd).

Poultry manure which is rich in nitrogen, part of which is volatile and so readily lost, is most useful to use when any organic matter is in a finely divided form. Such organic matter is usually known as 'screened dust' by town council experts. The fine organic matter improves the physical condition of the poultry manure and so enables it to be spread more easily and prevents excessive amounts of manure lodging in pockets with the consequent burning to plant roots.

It is suggested that poultry manure and finely divided organic matter such as Town Screening should be mixed together in equal parts. After this the temperature will rise considerably, and the heap can then be consolidated and allowed to mature for three weeks. The result is a dark uniformly blended material with an attractive earthy smell.

During the rise of temperature in the heap the weed seeds in the poultry manure are probably killed.

Pigeon manure is sometimes available from the lofts of racing pigeons or pigeon fanciers. Its value is double that of dried hen manure and it may be conserved and used as advised for poultry manure. It is very 'hot' and should therefore be used carefully, especially on soil intended for germinating seeds.

The guanos are really the residues of the excreta of fish-eating birds. They differ according to the weathering they have undergone, and those containing little nitrogen are usually much darker brown in colour, and have less odour than those which are poorer in this plant food. Curiously enough, if the nitrogen content is high the phosphate content is generally low, while the potassium content remains at between 3 and 6 per cent. Guanos are very quick in action, and should be regarded as about three times as rich as poultry manure.

ORGANIC FERTILIZERS

In addition to the organic manures, it is useful to be able to use organic fertilizers which give the plant foods needed. Organic fertilizers do not depress the humus content in any way, and undoubtedly help the physical condition of a soil.

FISH

Fish manure is very useful. It is equally valuable on light soils as on heavy soils. It is said, too, that the trace of iodine present is of value. Furthermore it may contain other trace elements. During its manufacture the oil is removed and a certain amount of potassium in an organic form is added. It is then comparable roughly to good guano and is used at the rate of about 135 g/m² (4 oz per sq yd). Worked into the top 90 or 100 mm (3 to 4 in) of soil, it is one of the best general organic fertilizers available. At the Arkley Horticultural College it is used in large quantities.

As I write this new edition it has become very difficult to get hold of.

DRIED BLOOD

Blood contains small quantities of potassium, sodium, calcium and phosphatic compounds in an available form. Its main function is, however, nitrogenous. It easily decomposes in soil, and is a plant food much beloved by glasshouse growers. It will often release its nitrogen as quickly as ammonium sulphate, and yet has the great advantage of being organic. It usually contains about 12 per cent nitrogen, and should be brought dry and powdery. Its application will be round about 65 g/m² (2 oz per sq yd).

HAIR, HOOF AND HORN

Sometimes it is possible to obtain quantities of hair from a barber's. This decomposes in the soil very slowly, supplying nitrogen over a period of five to six years. Feathers have a similar effect. The percentage of nitrogen will usually vary from 10 to 14 per cent.

The hoof and horn shavings from the blacksmith's can go into a similar category, and so can the leather wastes from leather factories. As a matter of fact, hoof and horn meal also contain 10 per cent of insoluble phosphoric acid, when it is mixed with a low grade of bone manure.

MEAT AND BONE

Meat and bone meal is popular, being made from slaughter-house offal. It contains from 3 to 8 per cent nitrogen and from 10 to 20 per cent insoluble phosphoric acid. Like hoof and horn meal mentioned above, it is often used in a similar way to fish manure, and is one of the ideal ways of applying nitrogen and phosphates in an organic form, because of the way they add humus, and improve the physical condition of soils.

BONES

The use of bone manure enables the gardener to apply

phosphates to the ground in an organic form. There are four ways of applying bones: (a) as bone meal; (b) as steamed bone flour; (c) as dissolved bones; and (d) as bone charcoal. In the case of bone meal only the fat is removed and the analysis works out at about 4 per cent nitrogen and 22 per cent insoluble phosphates. It is slow in action and so is of great value for use with slow-growing crops, for pot plants, or for crops of a permanent type. In the case of steamed bone flour the gelatine has been removed in addition to the fat and the analysis works out at about only 1 per cent nitrogen and up to 30 per cent of insoluble phosphoric acid. The phosphates, because of the fine grinding, are somewhat more quickly available and the product is easier to distribute because of its dryness. Dissolved bones are bones treated with sulphuric acid. The content of nitrogen is usually 2.5 per cent, of soluble phosphates 8 per cent, and insoluble phosphates 8.5 per cent. Phosphates are therefore almost immediately available, and for this reason some gardeners prefer it. Bone charcoal is used as a filtering medium when sugar is refined and usually contains about 34 per cent phosphoric acid, coupled with a somewhat larger quantity of lime. It is said to be of value on heavy soils.

WOOD ASHES

Wood ashes have been described as the ideal method of applying potash to the soils, for they help to reduce soil acidity because they contain much lime. The content of potash will vary from 5 to 15 per cent, the richest ashes being from burnt bracken and burnt bean haulm. It is usually applied at the rate of from 100 to 135 g/m² (3 to 4 oz per sq yd). Some gardeners, however, argue that it tends to make clay soils stickier – a fact, however, which the author has *never* been able to prove.

FLUE DUST

Flue dust, a material which comes from blast furnaces, may contain up to 15 per cent potash. The light buff-coloured dusts are usually richest in potash. Flue dust should always be applied in the winter or at least some time before sowing or planting a crop, because it may contain small amounts of substances such as fluorine which are injurious to crops. It gives very good results on acid soils.

Sedge peat is being used more and more in gardens today. It has the great advantage of being slightly alkaline, and can be applied to soil without fear of ruining its lime content. Gardeners, for instance, who have been most careful to see that the pH is absolutely right can use the best sedge peat with every confidence. It is, of course, pest and weed free, and thus is not like leaf mould, and when thoroughly soaked first, it does hold the moisture in a remarkable way and yet yields it up to the plant roots when they require it.

Peat is chiefly used in two ways. It can be applied with great success as a mulch. A 25 or 50 mm (1 or 2 in) thickness can be put on to the surface of the ground all over a shrub border, for instance, and so make hoeing unnecesary throughout the summer, and forking or digging undesirable, and unnecesary also, in winter. Each year further dressings of peat are applied in May if necessary, and much hoeing and forking labour is saved in consequence. The roots of the shrubs delight in a soil treated in this way and much better flowers and berries are assured.

Peat is often used with great success along the rows of strawberries, and it does save the use of straw for keeping the berries clean and, once again, is free from weed seeds. A strawberry plot with sedge peat remains almost weed free, whereas when straw is used, there is invariably a weed complex afterwards. Peat, in fact, is ideal to use along the rows of any plants and more and more mulchings of sedge peat will undoubtedly be carried out in the gardens of the future.

The second function of peat is concerned with providing the necessary fine organic matter and moisture-holding material for little seedlings. Well-rotted dung or compost may be forked in 50 mm (2 in) deep and is grand for plants when they have grown to a fair size and their roots have had a chance of reaching down. The seedling, however, needs some help in the early stages, and I have found that to rake sedge peat into the top 12.7 mm (0.5 in) of soil at the rate of half a bucketful per square metre (sq yd) does nothing but good, especially when, in the case of dry soils, it is thoroughly soaked first in clean water.

It will be realized that sedge peat is used in all the composts for plants to be grown in the greenhouse. It is grand for forking into greenhouse soils *in situ*, and whenever it is found difficult to obtain sufficient farmyard manure for compost, peat could be used at the rate of a bucketful per square metre (sq yd) or more. In this case, it will be necessary to use in addition, a 'complete' organic fertilizer, ie one containing nitrogen, phosphate and potash in the right proportions at, say 100 to 135 g/m² (3 to 4 oz per sq yd).

It may be as well to clarify the difference between sedge peat and sphagnum peat. Sphagnum peat is derived from the mosses and sedge peat from the sedges and rushes. In our original work on the subject in 1950-5 we discovered that these two peats were quite different, both in composition and value, and that sedge peat always gave us better results.

We have on a placard at our Open Days at Arkley Manor Gardens expressed the comparison of the two peats as follows:

COMPARISON OF PEATS

	Sedge peat	Sphagnum peat
Available humus Nitrogen	100 kg (224 lb)	20 kg (45 lb)
Additional humus	63.5 kg (140 lb)	21 kg (46 lb)
Protein	70%	15%
Acidity	high	little or none
	pH5.0	pH4.0

This gives quite a good picture as to the value of the one peat versus the other and does show the importance of buying the right peat for the job.

As to the use of sedge peat in the garden, it is undoubtedly advisable to use the medium grade. The very coarse type gets taken by the birds to make nests and the very fine peat easily gets blown off the beds by the wind.

When peat is used as a mulch it should go on at 25 mm (1 in) deep all over the bed. This means that 50.80 kg (1 cwt or 1.12 US cwt) of medium-grade sedge peat will cover a bed of about 9 m² (11 sq yd).

Mulching methods Use medium-grade sedge peat all over the beds which are growing permanent crops and semi-permanent crops. I refer to such beds as those for roses, heathers, flowering shrubs, irises, primulas, carnations and pinks, rhododendrons and azaleas. You can also use the sedge peat for strawberries, raspberries and other soft fruits. It is excellent as a mulch for the cut-flower border, the rows of more permanent vegetables like asparagus, seakale, artichokes, as well as for such crops as tomatoes, cucumbers, marrows, squashes, celeriac, peas, beans and parsley.

In the case of such permanent beds like roses and shrubs, it is perhaps best to make the initial application in May after the snow has gone and the frost has come out of the soil. It is possible, however, to apply the peat at any time, but if you do cover frozen soil with organic matter then the tendency is to keep it cold, ie it will warm up much more slowly. I prefer, therefore, to start this ideal mulching scheme in the spring, or to apply the peat in the autumn before the onset of frosts.

What mulching does When the sedge peat mulch is put all over the bed, the annual weed cannot grow. Thus there is no hoeing or forking to do for the life of the bed concerned. At the Arkley Horticultural College there are rose beds – a dozen or more in all – in a large rose garden which had not been hoed for twelve years. The sedge peat mulch made this quite unnecessary.

You cannot, of course, smother and prevent perennial weeds from growing by mulching with sedge peat. So if you

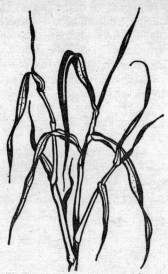

Couch or twitch grass.

are unlucky enough to have a garden with thistles, nettles, docks, couch or twitch grass, ground elder and the like, you must get rid of these with a strong hormone weedkiller. It is necessary to start with a garden free of perennial weeds, and then to use the sedge peat to control the annual weeds.

It must be remembered that every time a gardener hoes his soil he disturbs thousands of weed seeds that have been lying dormant. This happens inevitably when you dig or fork. Move soil and you unfortunately disturb the weed seeds that may have been lying there for years and years, and they start to grow and have to be controlled. The mulching scheme makes soil disturbance quite unnecessary.

What the worms do The worms love sedge peat and they start to increase and work the moment a layer of this organic matter is applied as a top dressing. This is a good thing because they do the necessary cultivation, often tunnelling down 0.6 (2) and 0.9 m (3 ft) and sometimes even up to 1.8m (6 ft). This causes perfect aeration, as it lets the air get down into the soil. It allows the moisture to seep down as well as plant foods.

I find that the first year the worms pull in about 6 mm (0.25 in) in depth and so I top up with 6 mm (0.25 in) sedge peat in the spring of the following year. This usually happens the second and third years too and so I repeat the 'dose' each time. After three or four years the worms have usually pulled in what they need to help build up the humus content of the soil and so the sedge peat remains as a weed-smothering blanket all over the soil and hardly ever needs to be disturbed.

Feeding and sedge peat It is usually necessary to feed the plants where sedge peat is being used as a mulch and in this case fish manure or seaweed manure may be used at about 85 (3) to 113 g (4 oz) per m² (sq yd). This is applied as a top dressing, being sprinkled on in February as a rule, and is allowed to work through the peat gradually. If a second application is deemed advisable because growth isn't quite as desired, then this may be given in late July.

It is seldom that one needs to give fish manure to a shrub border, but it is invariably necessary in the case of rose beds, strawberries, raspberries and other soft fruits. Once a year I use fish manure or a suitable substitute also on the herbaceous and cut-flower borders.

Lime and sedge peat When I started using sedge peat as a mulch the 'experts' said that I couldn't apply it on special beds devoted to irises or the carnation or pink family, for instance. Peat, I was told, was acid and would be hated by plants that liked lime! Actually this is only a half-truth.

I get over any troubles over lime by scraping away with my hand the dry sedge peat around the plant that loves lime, and put on the soil I expose in this way a dessert-spoonful of hydrated lime. I then put back the peat over the top and all is well. The lime is on the soil below and soon washes in; the plant is happy; and the sedge peat, going as it does back into position again, leaves the bed tidy and neat.

Sedge peat and beauty   People who come (as they do in large numbers) to see The Good Gardeners' Association demonstration gardens at Arkley Manor, near Barnet, Hertfordshire, are always surprised to see the beds looking perfectly tidy and not a trace of weed to be seen. They are astonished at the lovely dark brown of the soil – as they think it is. The soil, actually, is horrible – just gravel over London clay – so when you see it *au naturel* it's full of stones and looks as poor as Lazarus! Cover it with sedge peat and it looks terribly attractive.

So with sedge peat not only are the annual weeds smothered, the moisture kept in the ground, the soil kept warm, but, in addition, the garden looks beautiful.

CHAPTER 7

Composting Methods and
Composting Instructions

Because of the lack of all kinds of animal manures, and because men and women have seen the disastrous results all over the world of not adding organic matter to the soil regularly, there has been a great national swing to the composting of vegetable refuse. No wise man today burns his rubbish (although some British Royal Park keepers still burn leaves). The wealth of soil-feeding materials found in the unwanted parts of vegetables and of plants such as weeds, leaves and grass mowings, should be conserved and used, and not dissipated into thin air. It is impossible to grow healthy vigorous plants without humus, and humus has been described as 'vegetable "mould" manufactured from mixed plant wastes by living organisms'.

It is obvious that the ideal method of composting vegetable refuse should be as near as possible to that of Nature. In the forest, for instance, there are layers of decaying leaves and most of them have some manure on them, say that of bird, beast or insect. These are under the trees often piled to a height of 0.9 m (3 ft) and this protection from rain and wind enables the fungi and bacteria to work well. The rotting is done in the main by the fungi, helped by manurial activators and continued by the bacteria. In the early stages there is a fair amount of heat generated and the temperature may rise to 82 °C (180 °F). The speed of rotting depends on the nature of the material, the moisture content, the conservation of the heat, the degree of aeration and the method employed for counteracting acidity.

There are various methods advised by experts on the subject of composting. All these have, however, certain points of agreement, and it is these salient features which must first of all be emphasized.

1 The layers of plant refuse used should be about 150 to 210 mm (6 to 9 in) in depth, and should be trodden down moderately firmly. Occasional layers of farmyard manure 50 mm (2 in) in depth may be included.

2 A shelter should be provided from wind and sun to obviate drying out and loss of heat.

3 The material as far as possible should be used when fresh and juicy. Very fibrous material such as the stalks of cabbages should be crushed on a chopping block with the back of an axe or cut up finely with a spade. Such material is always better intermingled with soft matter such as grass mowings and the like. Very woody material such as winter apple prunings should be excluded. This should be burnt and the ashes added to the heap.

4 The heaps should never be too large or too high. Maximum height 2.1 m (7 ft). Garden heaps may be 1.2 m (4 ft) by 1.2 m (4 ft) to 3 m (10 ft) by 3 m (10 ft).

5 A certain amount of lime may be used, say 300 g of hydrated lime per square metre (1 oz per sq ft), every 300 or 600 mm (12 or 24 in), if the garden soil is very acid, but as a rule only for the first two or three years.

6 Some kind of accelerator is necessary. Experts advise organic manure, poultry droppings, rabbit manure, fish manure, farmyard manure – and of course seaweed manure.

One of the great difficulties of the Indore method (p 76) is the amount of farmyard manure used as an accelerator, for this is almost impossible to obtain except when living near farms.

It is obvious that in addition to these numbered details there are many other points on which all composting experts would agree. For instance, there is no need to shake the soil off the roots of weeds when throwing them on to the heap. It is a good plan to start a heap in the early summer which can be ready for use in the autumn and another heap in the autumn for use in the spring. If fresh greenstuff is used that seems to ensure rapid decomposition and so obviates the need for any turning over.

It is important to see that a high temperature is engendered because then the weed seeds and diseases will be killed. No one should be discouraged by the seemingly conflicting instructions, and there is no doubt that any good method of composting vegetable refuse is well worth while, whether such composting takes six months or a whole year. The great thing is to compost.

When the material is friable, sweet, powdery and dark brown or black in colour, it is ready for use. There is no need to dig it deeply. Far better use can be made of it if it is worked or scuffed into the top 50 or 75 mm (2 or 3 in) of soil. The tilth will thus be improved, the soil will not dry out so readily and there will be ample humus for feeding the seedling plants as they develop.

Chapter 8 deals with green manuring, but there is no reason why the keen composter shouldn't grow certain crops like mustard, lupins, field peas, vetches, rye-grass, etc, for the purpose of cutting before there are signs of seeds forming and then carting on to the compost heap There is little doubt that better results will be achieved with using such crops composted, rather than digging them into the ground where they grow.

IDEAL HEAP BUILDING

It has already been made clear that the heap should be built layer upon layer. The sappy and finer substances should be about 150 mm (6 in) in thickness; the more bulky and fibrous substances can be in slightly thicker layers. The aims however, should be to have the layers alternately, ie bulky material and fine material, and dry matter next to wet substances. If care is taken in the build-up much better rotting takes place and the final compost is far more uniform.

In between the layers of vegetable waste should be placed the accelerator. In The Good Gardeners' Association method, this will be a good fish manure, a seaweed product or dried poultry manure. An organic liquid manure is also good.

The height of the heap will vary somewhat according to

71

the size of its base, but normally it should be 1.8 or 2.1 m (6 or 7 ft). It is convenient (unless the material has been kept in a wooden bin) to allow the heap to slope slightly inwards, then to have a flat top which makes it convenient for watering. Such a heap is easy to cover with soil in a similar manner to a potato clamp. If the heap is made outside it is useful to keep it covered with old sacks or to use bracken or straw which will help to conserve the heat.

When a heap is built gradually, day by day and week by week, it is seldom necessary to tread the material to consolidate it, although personally I tread each layer once every evening especially around the sides of the bin. On the other hand, if because of the large amount of vegetable refuse available at any time, the heap can be made on any one day, much treading will be advisable, especially if it is composed largely of bulky material. In this case the heap will be covered with soil on the same day for the purpose of retaining the heat and moisture. It should be 75 mm (3 in) thick so as to be protective but not impervious.

It is better to use a good loam than a sticky clay or gritty sand. If it is impossible to get anything else than the latter the thickness will have to be increased to about 150 mm (6 in), or, if the former, reduced to only 50 mm (2 in).

One of the disadvantages of covering a quickly made heap with soil is that sinking will take place during the rotting-down process. Some gardeners therefore prefer to leave the heap for a fortnight before the earthy cover is put on. In this case some temporary coverage of old sacking or matting is usually employed.

MANAGING THE COMPOST HEAP

Most heaps rot satisfactorily without any attention at all. When ready to use, the compost should look like a powdery earthy mould and should not show traces of the original materials. It is curious the way that earthworms leave the heap as it becomes mature. However, there is nothing against using partially rotted compost for mulching crops or for potting, for in these cases the rough

72

portions prove useful. A compost heap need not be used for about two years, for during this period it tends to improve, but after this time there is likely to be a loss of plant foods.

It is interesting to have an occasional inspection of the interior of the heap by delving in with a good trowel. If the centre has a pleasant earthy smell and feels like a moist sponge, though when squeezed does not ooze out water, then nothing need be done. If it is wet, sour smelling and slimy, it is worth while removing the covering and turning the heap to mix the wet and dry portions together. At the same time, a little more soil could be given. After turning, wet sacks should be placed over the heap and it should be allowed to settle for a week before the soil covering is put on once more.

If the examination shows a whitish growth, brittleness plus a musty smell, and often the presence of wood lice, then moisture is indicated. The trouble can often be overcome by the use of dilute liquid manure or bedroom slops. Such a heap can be wonderfully improved if turned in the open during a rainy period. During the last fourteen years I have never had to turn a heap.

USING THE COMPOST HEAP INTELLIGENTLY

It has already been suggested that compost should not be buried deeply and gives the best results when mixed thoroughly by scuffing into the top 50 or 75 mm (2 or 3 in) of soil. If it is more convenient to dig it in, this again should be done shallowly so that the material is not buried deeper than a 75 or 100 mm (3 to 4 in) depth. The aim should be to apply compost little and often. It gives very good results when worked into the ground before the sowing or planting out of every crop, no matter what period of the year.

Those who have large quantities and large gardens should use anything from 180 to 250 kg per 100 m² (3 to 4 cwt or 3.36 to 4.48 US cwt per 100 sq yd), while in the small garden, quantities will vary from 1.6 to 2.1 kg/m² (3 to 4 lb per sq yd). The higher figures should always be aimed at, though if

73

more than one application is to be made to a particular plot of ground in any one year, 2.1 kg/m² (4 lb per sq yd) on each occasion would ensure very good results.

THE GOOD GARDENERS' ASSOCIATION METHOD (SHEWELL-COOPER)

For every 150 mm (6 in) thickness of vegetable refuse collected, a good fish manure or seaweed manure is sprinkled on at 100 g/m² (3 oz per sq yd), another 150 mm (6 in) thickness should then be placed in position, and yet another sprinkling of fish manure and so on. It may take days to reach the 150 mm (6 in) thickness, or because some crop like potatoes is being harvested it may be arrived at in one day or sooner.

Where it is possible to obtain the urine from animals or the excreta from poultry or rabbits it can go on the heap in addition to the fish manure. Once a fortnight or so if the weather is very dry, the heap may be given a good watering and when it gets 0.9 or 1.2 m (3 or 4 ft) high a fork may be plunged into it at two or three places. Such vegetable waste would be ready for use in about six months. It is convenient therefore to have two heaps going – each 1.8 m by 1.8 m (6 ft × 6 ft), one which is almost ready to use, and the other which is being piled up to use in six months' time.

If the garden soil is known to be acid, a sprinkling of hydrated lime is given every two feet at the rate of 135 g/m² (4 oz per sq yd), instead of fish manure or poultry manure layer. This is unnecessary after the first three years.

The material should be kept in a bin made of wooden planks leaving an inch between the planks to let in air. Such a heap controlled on three sides rots down extraordinarily well and never gives any trouble from being over-wet.

TEN COMPOST COMMANDMENTS (GGA RULES)

1   The bin must be square, say 1.2 m by 1.2 m (4 ft × 4 ft), 1.8 m by 1.8 m (6 ft × 6 ft), or 3 m by 3 m (10 ft × 10 ft).

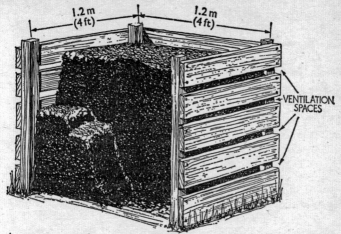

A square, wooden, well-ventilated bin. Ready for use at the end of six months, some of the brownish-black powdery compost has been removed.

2 It must be made of wood.

3 The planking used must have 25 mm (1 in) spaces between them.

4 The heap must be made on soil so that the worms can get up in the heap.

5 The layers of vegetable waste must be 150 mm (6 in) thick.

6 Each layer should be sprinkled with activator at 100 g/m² (3 oz per sq yd).

7 Every night before going to bed the vegetable waste should be trodden level, particularly along the sides.

8 If great quantities of lawn mowings are used, put on a 50 mm (2 in) layer and lay a sheet or two of newspaper on the layer, then another 50 mm (2 in) of lawn mowings and so on. (This ensures that it doesn't turn into a soggy mess.)

9 The wood must be treated with Rentokil preserving fluid and NOT creosote tar or solignum (our bins at Arkley have lasted thirteen years so far).

10 When the heap has risen to 1.8 m (6 ft) tall, cover the
    top with a layer of soil 50 mm (2 in) deep or use
    several thicknesses of old blankets or old rugs.

NB If when you are treading the material in the bins level
in the evening you urinate on the heap, it will do a great deal
of good.

It is possible to make excellent compost without turning
the heap at all.

THE INDORE METHOD (SIR ALBERT HOWARD)

The late Sir Albert Howard in a broadcast during the last
war appealed for a 'Grow Better Food Campaign'. He suggested that for the small garden a heap, 1.2 m by 1.2 m (4 ft
× 4 ft) and 1 m (3 ft 4 in) high, should be aimed at. In such
an area 1.5 m³ (2 cu yd) of compost could be made weighing about a tonne (1 ton or 1.12 US ton). To make the box
to contain this compost as it is rotting down, six lengths,
1 m (3 ft 4 in) long, of 50 mm (2 in) by 50 mm (2 in), were
needed for the uprights, and twenty-four 1.2 m (4 ft) lengths
of 150 mm (6 in) by 25 mm (1 in) boards for the four sides.
This timber should be oiled with old motor oil to preserve it
and should not be planed.

Six of the boards should be nailed to two of the uprights
to make one side, leaving a 12.7 mm (0.5 in) gap between
the boards for ventilation. This should be repeated for the
other side and the end. The three sides of the bottomless
box can then be bolted together by means of four bolts
while the front of the box should be made up of loose
boards, 150 mm (6 in) by 25 mm (1 in), slipped behind the
uprights as the box is being filled. It is a good plan to use a
wooden bar, 50 mm (2 in) by 36 mm (1.5 in), fitted with two
wooden blocks, 76 mm (3 in) by 50 mm (2 in) by 280 mm
(11 in), across the top of the box to prevent the sides spreading outwards.

The mixed withered vegetable waste, cut up into lengths
a few inches long if possible, should be placed into the box
together with one-third to one-quarter of the same volume

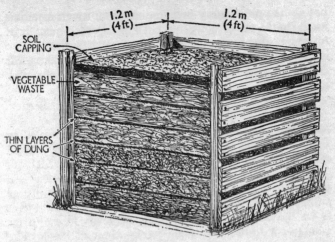

The Indore Method. The activator used here is some form of dung.

of manure. A little soil should be incorporated at the same time. If there isn't any pig, poultry or rabbit manure available, substitutes like hoof and horn meal, fish manure, or dried blood should be used instead, at the rate of 1 to 2 per cent of the dry vegetable waste. Where these substitutes are not available, bedroom slops may be used.

When the box is full, two or three vertical holes should be made in the mass to improve the supply of air, and the top should be covered with two pieces of old thick blankets to keep out excessive rain. Active fermentation should cause the mass to begin to sink and it is then that the vacant space should be filled up until no further subsidence takes place. After six weeks the material in the box should be dug out and stacked nearby on a convenient site, where it should be allowed to ripen for a month or six weeks. Here it should be watered if necessary to keep damp. Sir Albert estimated that at least 4 tonnes (4 tons or 4.48 US tons) of compost can be made per year in one of these 1 tonne boxes.

The main points of Sir Albert Howard's scheme seem to

be: (a) the preparatory breaking up of all rough materials used; (b) the use of powdered chalk rather than quicklime; (c) composting in pits or boxes; (d) efficient turning; (e) the inclusion of some animal wastes. Generally speaking Sir Albert advised two turnings of such heaps at four-week intervals.

THE Q R METHOD (MAYE E. BRUCE)

Miss Maye E. Bruce, who at one time was connected in Britain with the Anthroposophical Foundation, developed what is now known as the Q R or Quick Return Method. This has many devotees because good-quality compost can be produced in her way in the minimum time. It is just as simple and effective for the small garden or allotment as it is for the larger estate. The heaps are ready to use in the summer in from six to eight weeks' time, and only take about twelve weeks in the winter. A herbal solution is used, consisting of chamomile, dandelion, valerian, yarrow and nettle, plus powdered oak bark and pure run honey. The herbs are properly dried and powdered, and though the formula is not secret, most people find it convenient to buy the activator ready-made. It is presumed that this activator works as a catalyst and it is used of course in what may be called homeopathic doses. The powder is usually soaked in rainwater for some hours, and the liquid is then poured into holes made in the heap.

The Quick Return Method has been used with equal success on clay, on rich valley soil and on sand. By making use of the natural heat of decomposition and by retaining it, the rotting-down process is hastened. 'The vital forces of plant growth are free,' wrote Miss Maye Bruce. 'If these can be kept in the heap they radiate through it, charging the resultant soil with the vitality of plant life, revitalizing it, restrengthening it, and through it returning to the earth the life that the plants have borrowed for their fruition.'

MAKING THE BIN

The success of this scheme depends on several factors:

(a) the use of a bin of some kind, preferably of wood; (b) the correct foundation of the heap; (c) the correct building of the heap; (d) the cover to keep the rain out in wet districts; (e) a warmish aspect. Naturally, the size of the bin depends on the size of the garden and the amount of material it is proposed to compost during the year. For a small garden, a packing-case 0.6 m (2 ft) square will do. For a medium-sized garden a bin 0.9 m (3 ft) by 0.9 m (3 ft) by 0.9 m (3 ft) made from old planks will probably be suitable. The outside cuts of trees, which timber merchants usually call slabs, are often released from sawmills for this purpose.

It is a good plan to have more than one bin so that while one heap is rotting down another heap may be made. If wood is impossible to obtain, a good expedient might be the use of baled straw walls, while for those who find it difficult to get baled wheat or barley straw, special corrugated cardboard can be used, tacked on to stakes or woven in and out of a number of stakes so that it gets support on both sides. Three-ply wood should not be used as it is impervious to air, nor should corrugated iron. Concrete or brick are not advised as they do not admit sufficient air, and aeration is of vital importance. Corrugated iron, however, may be used as a roof to keep off rain, and so may other convenient materials, such as doubled old cement bags, or old sacking hung taut like a tent ridge.

THE CORRECT FOUNDATION

The bottom of the bin should always be the earth, never concrete or brickwork. If the soil should be very sticky a little loose rubble may be worked in to ensure good drainage. A sprinkling of charcoal at the bottom will help keep a heap sweet.

BUILDING THE HEAP

Place in the bin on the foundation all vegetable refuse, with the exception of the autumn leaves and dry sticks. Do not put on to the heap any paper, animal remains, metal, skin, flesh, fat, bones or any left-overs from the kitchen.

Miss Bruce claimed that Nature does not use decaying animal matter as food for plants. The layers of material should be kept flat, and should be trodden down gently to prevent air pockets. Any material which is long should be roughly chopped into lengths of about 0.3 m (1 ft) with a sharp spade. Alternate fresh juicy stuff with coarser material, making each layer about 100 to 150 mm (4 to 6 in) thick. Hedge prunings are good only if pliable and green. Grass mowings should be used in layers not more than 100 mm (4 in) thick.

A thin layer of manure will accelerate the breaking-down process, though it is by no means necessary. Occasional spadefuls of soil should be added and two light dressings of lime during the making of the heap at, say, 0.3 m (12 in) and 0.6 m (24 in). The lime should be separated from the manure by a layer of greenstuff, or a chemical action will be set up and some of the nitrogen lost from the heap. Quicklime or ground-lime is best and this should be kept in an airtight tin to prevent it slaking. It should be applied at the rate of ten 12.7 mm (0.5 in) well-pulverized cubes per square metre (or one cube per square foot). If quicklime cannot be bought, hydrated lime should always be used at approximately twice the rate.

A sack should always be placed directly on the compost heap. It keeps the damp heat in and prevents the sun and wind from drying out the top layer. In this way the weak seeds are killed. When the bin is filled and firm, the heap should be covered with an inch of soil. No turning should be done.

The heap is treated as it is being built by sprinkling each layer with Quick Return solution. The Quick Return material itself is a dry powder and a level teaspoonful is mixed with 0.6 l (1 pint) of water, shaken up well and allowed to stand for a few hours before use. A volume of 0.6 l (1 pint) of solution is enough for a heap up to 0.9 m (3 ft) by 0.9 m (3 ft) by 0.9 m (3 ft), and each layer, as it is built, should be sprinkled with the liquid. There is no secret about the herbs which are fully described in Miss Bruce's book *Common-Sense Compost Making*.

When ready the compost should smell sweet and should be easily broken up by the fork. The quicker the heap is built the sooner it will be ready. Try to get each heap built within six weeks. Dig into the heap with a trowel and sweet soil should be found. If the material smells strong and is slimy, then the heap must be left longer. If it appears sweet but is too closely packed and so not all humus, loosen it so as to let the air get at it and it will break down quickly.

In the spring and early summer the heap will be ready in between four and eight weeks, in the summer and early autumn between ten and twelve weeks, and in late autumn between three and four months.

These, Miss Bruce considered, had lost their vitality and had no heat to give. They should therefore be stored in a heap by themselves with soil in layers, 150 mm (6 in) of leaves and 25 mm (1 in) of soil, and be left for six months. The heap should then be turned and stacked again, and as a result a rich black soil will be produced.

NB re Q R

I have tried this method again and again but I find it only works when soft organic waste is used. It is useless on cabbage stumps, the tops of herbaceous plants and such like.

SPECIAL NOTES AND THOUGHTS ON COMPOSTING AS A WHOLE

1 A properly made compost heap is not a smelly rotten rubbish heap.
2 The work of the bacteria in the heap may cause a gain

of 25 per cent more nitrogen than actually went into the heap either as waste or as activator.

3 The reason lime is used in the early stages of compost gardening in the heap is that the bacteria are most active between the pH7 and pH8.

4 After three years the compost gardener may be able to do without using lime, because the earth which gets on to the heap on the roots of the plants and weeds will do the trick.

5 The temperature of the heap may easily rise to 82 °C (180 °F) and it is then that the actinomycetes can break down the more resistant carbohydrates and proteins.

6 The temperature may remain up to 49 °C (120 °F) for a month, and then the 'hot loving' organisms are replaced by the bacteria which prefer normal temperatures. These attack the disease-producing organisms, while some fungi

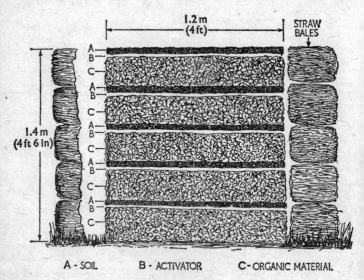

A - SOIL   B - ACTIVATOR   C - ORGANIC MATERIAL

A compost bin made of straw bales and in which layers of soil and activator are placed between layers of vegetable waste.

82

feed on the eelworms. It is at this stage that the antibiotics are produced, as well as the hormones or growth-promoting substances, and it is these that undoubtedly make compost-grown plants more disease resistant.

7   Compost should always be used when it is mature. It is then a dark-brown or black powder. Sweet smelling, it looks like moist peat, and 85 per cent of it will pass through a 12.7 mm (0.5 in) sieve.

8   Don't look on compost as merely a substitute for farmyard manure. When properly made it is three times as valuable as old dung.

9   Its analysis may well be 5.5 soluble humus, 22.1 organic matter, 0.9 nitrogen (N), 0.7 phosphates (P), and 5 per cent potash (K). In addition, however, to the actual NPK plant foods it contains, it is 'alive' with billions of micro-organisms per gramme. It will contain also what are usually known as the minor minerals or trace elements (micro-nutrients) which the plants require, *and* the invaluable antibiotics, enzymes and vitamins.

10   Compost contains micro-organisms capable of producing antibiotics. Such micro-organisms colonize the soil around the plants and produce antibiotics continuously. Antibiotics affect the micro-flora of the soil. They can be taken up by the root system and exert a systemic effect preventing infection in the stem and leaves.

Most active antibiotics are chemically sound and cannot be recovered from soil, if added. If however the antibiotics are produced in continuously close proximity to the roots, it isn't necessary for them to survive in the soil for long periods of time.

CHAPTER 8

Green Manuring and Sheet Composting

Green manure comes from plants cut down before maturity, smashed up with a spade, and dug in. A great deal of unfortunate advice has been given about it. It has been said, for instance, that it can be the ideal substitute for manuring with compost or with well-rotted dung. This is not the case. Many green manures cause nitrogen starvation when dug into the ground since the bacteria have to borrow nitrogen from the soil in order to rot down the stems and leaves that have been dug in. Thus the raw organic matter proves a big drain on the biological soil energies. Contrary to popular belief, fresh green manure will in fact cause a reduction in the humus content of the soil. What happens is that the organisms attach to the green matter, and decompose it, and then because there is an excess of nitrogen the bacteria proceed to use up other more resistant organic matter which may be present in the soil. The growth of plants sown on a green-manured piece of land may well be retarded during the rotting-down process of the leaves and stems. There is always a great deal of acidity produced during decay, and this, too, is injurious to most plants.

There are certain exceptions to this, and one, of course, is the annual lupin. This perhaps is the reason why the Germans use it extensively. The lupin has nitrogenous nodules on its roots, and so when the plants are dug in, there is no serious nitrogen robbing. One problem, however, is that undigested organic substances may affect the roots of plants adversely.

It has been claimed that the absorption of partially decomposed nutrients from the proteins may actually have an effect on the nature of the sap in the plants, with the result that the leaves and stems are more subject to pests and disease.

Green manures, therefore, can never be relied on to produce as good results as properly composted vegetable waste. They are in fact very slow in action. It often takes six months – especially in the winter – before the green matter has rotted down into a condition suitable for the plant roots to feed on. A green crop may render a soil too dry, unless means are taken to supply adequate moisture during the growing season.

Spear thistle.

The biological factor plays the greatest part in the success of green manuring. The micro-organisms cannot work unless they are given the right conditions. The soil, therefore, that is to be green manured must be perfectly drained so that there is ample air. There must be sufficient lime in the soil to prevent acidity, and sufficient nitrogen present to provide the energy with which the organisms can start their work. The soil too must be warm. These conditions are usually found only in the more temperate climates where the rainfall is well distributed and where the soil is ideal.

DOUBLE GREEN MANURING

Mr R. L. Scarlet, VMH, CDA, of Musselburgh, Scotland, has worked out a system of double green manuring designed to obviate the difficulties over denitrifications. His scheme also plans to ensure the total destruction of docks, couch grass and other perennial weeds. The remains of these weeds last much longer in the soil than the sown green manures, and so they themselves form a valuable source of humus, giving back to the soil all that has been extracted during their growth. The accumulated humus thus formed not only helps to hold the food material, but it also increases the soil fertility in a remarkable manner.

The land is dug during the winter and early spring, and tares are sown broadcast at the rate of 17 g/m² (0.5 oz per sq yd). When these come into flower late in June, the plants are smashed up with a spade, and nitro-chalk is applied at the rate of 34 g/m² (1 oz per sq yd). (The author prefers to use dried blood or dried poultry manure at 68 to 100 g/m² (3 to 4 oz per sq yd).

When this has been done, material should be left for eight days, and should then be shallowly dug in. The surface of the soil should then be raked down so that rye may be sown at 34 g/m² (1 oz per sq yd) during the third week of July. This rye is then knocked down with a spade, any time between October and December, and is dug in shallowly also. This method of double green manuring does enrich the land.

Gardeners must note that a whole year must be given to this planning, and the result is, of course, that one cannot crop the particular area with vegetables. Very dirty ground can, however, be cleaned in this way, and humus added. The scheme may recommend itself heartily to some readers.

Market gardeners sometimes sow a grass mixture over the soil in the spring with the idea of producing what is called a temporary ley. The grass thus produced is either grazed with goats, sheep, geese or poultry, or is kept cut every ten days, the grass clippings being allowed to fall back to the soil. At the end of the year this grass is dug in like manure. Before it is incorporated, however, a fish manure may be applied all over the surface at 100 g/m^2 (3 oz per sq yd) or dry poultry manure used at 132 to 170 g/m^2 (4 to 5 oz per sq yd). If the temporary ley has been kept short by means of animals or birds, it is far richer in natural nitrogen than when it has to be mown.

Gardeners may like to try this temporary ley system, and if so they should obtain from the seedsman a ley mixture consisting of four parts perennial ryegrass S.23, four parts timothy S.51, one part clover S.100, two parts cocksfoot S.143 and two parts of rough-stalked meadowgrass, at 17 to 34 g/m^2 (0.5 to 1 oz per sq yd).

CHOOSING THE CROP

Reference has already been made to the various crops that may be used in green manuring. Many gardeners find it economical to sow any seeds they may have left over at the end of the gardening year or save them for sowing the following season. The broadcast sowing of mixed seed has its advantages and its disadvantages, but it certainly is a way of producing green matter for digging-in at very little expense.

Mustard has the advantage of being a very quick grower. It can usually be dug in at the end of a 7-week period. On the other hand it is subject to the club root disease so should never be used on land which has been affected by this trouble. It is a very suitable crop for the light sandy or gravelly soil. Other crops for this type of land include tye and lupins.

It takes as much as 8000 kg of carefully preserved poultry

manure per hectare (64 cwt or 72 US cwt per acre) to compost *in situ* mustard which has been sown as a green manure. Composting on the plot where the vegetables are to be grown, instead of in the heap, is sometimes called sheet composting.

For the medium loams there are a number of crops which have given good results. They include rape, rye, field peas, vetches, various types of clover and lupins, while for the heavier soils such as clays, vetches and red clovers seem to be the right 'medicine'. Experiments however are still being continued on the subject of green manuring.

QUANTITIES TO SOW

It is difficult to lay down hard-and-fast rules as to the quantities of seed to sow for manuring, but the following figures should act as a guide:

Mustard: 2 g/m^2 (0.0625 oz per sq yd) or about 15.7 to 22.5 kg/hectare (14 to 20 lb per acre).

Vetches: 2 to 4 g/m^2 (0.0625 to 0.125 oz per sq yd) or about 22.5 to 31.4 kg/hectare (20 to 28 lb per acre).

Lupins: 4 g/m^2 (0.125 oz per sq yd) or about 90 litres/hectare (2 bushels per acre).

Clovers: 1 to 2 g/m^2 (0.0312 to 0.0625 oz per sq yd) or about 15.7 kg/hectare (14 lb per acre).

Oats: 8 to 25 g/m^2 (0.25 to 0.75 oz per sq yd) or about 135 to 180 litres/hectare (3 to 4 bushels per acre).

Field peas: 34 g/m^2 (1 oz per sq yd) or about 160 to 190 kg/hectare (1.25 to 1.5 cwt per acre).

Green manuring will always prove a most convenient means of adding humus to ground where, because of difficulties of access or for some other transportation reason, it is difficult to get large quantities of compost or old rotted farmyard manure on to the ground.

SPECIAL NOTES AND THOUGHTS ON GREEN MANURING

1   One can ensure that there is the right amount of moisture in the soil, when green manuring, by using overhead irrigation by means of, say, a square-area rainer or whirling spray.

2   Diluted liquid manure is sometimes used over green manure before digging it in.

3   Do the digging-in while the soil is warm, in the summer or autumn.

4   The leguminous plants which collect nitrogen from the air through the bacteria which live in symbiosis on the roots are lupins, vetches, peas, and clovers.

5   Where wireworms abound use mustard alone, sowing it at the rate of 68 g/m² (2 oz per sq yd).

6   Fresh green matter put into the drills at potato-planting time does prevent an attack of potato scab.

CHAPTER 9

Lime – Acidity and Sourness

The gardener must aim to keep the acidity of his compost heap and his land about neutral. If the land is too acid, it may be necessary to add lime in some form. If too much lime is applied then some of the minerals in the soil, especially iron, will be locked up, possibly causing plants to suffer from what is called lime-induced chlorosis. Most gardeners find it convenient to use lime as a form of neutralizer, though in parts of the world where it is not easily available a mixture of wood ash, urine and earth may be used instead. The farmers of years gone by used to use marly clay, because it contained approximately 10 per cent of calcium carbonate.

Soil acidity and soil sourness have been described as synonymous terms, but they are not really the same. Unfortunately, what the gardener means when he calls his soil sour is not easy to explain. Sourness is common in town gardens as well as in some allotments in various parts of the country. It is often coupled with dankness, with darkness and with incomplete decay. Mosses and other green lowly algae seem to flourish on the surface of sour soil. Sour land is often badly drained, and it is one of the shameful things of the past that when new estates have been developed on well-drained land, the agricultural drains have been blocked and broken during the process of building, resulting in ill-drained gardens and even flooding in gardens which lie lower than others.

Most of the ditches in the country need digging out so that the outfall of the drain pipes can be carried away quickly. The removal of drainage water in this way might do much to obviate sourness, for the stagnant stinking water that collects in the subsoil may be largely the cause of the trouble. In ill-drained land the natural processes of decay

THE IMPORTANCE OF DRAINING SOIL

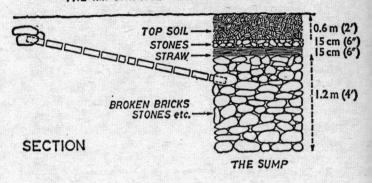

TOP SOIL → 0.6 m (2')
STONES → 15 cm (6")
STRAW → 15 cm (6")

BROKEN BRICKS
STONES etc. → 1.2 m (4')

SECTION

THE SUMP

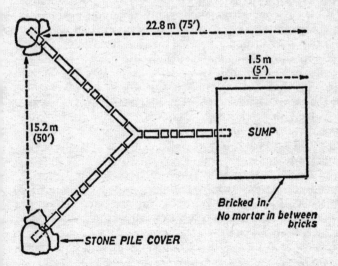

22.8 m (75')

1.5 m
(5')

15.2 m
(50')

SUMP

Bricked in.
No mortar in between
bricks

STONE PILE COVER →

PLAN

The sump which can accept drainage water from the soil. Note the agricultural drains.

are arrested because the air-loving bacteria cannot work. The result is usually an accumulation of foul liquor which may give off gases to spoil the soil above. When there is free drainage the decaying of vegetable matter can be completed without sourness or foul smell.

Sourness is often seen in grazing pastures shaded by hedges or tall trees. One of the main causes of the trouble, however, is lack of lime. The application of lime, therefore, can do much to prevent both acidity and sourness.

It has been said that any soil containing less than one part of lime in two hundred parts of soil cannot be brought into a high state of fertility. Liming does correct undesirable soil conditions. Although it must be regarded principally as a soil-improving agent, it does, however, provide the calcium needed as a food by plants, and improves the texture of heavy soils by making them more friable and workable.

Lime, by counteracting the acidity in the soil, enables the micro-organisms (whose activities are associated with fertility) to work in an unrestricted manner. It ensures the right decomposition of organic matter. It also renders soil phosphates more available. It has a neutralizing effect on the acids in peaty soils, the presence of which is unfavourable to the growth of desirable plants. It helps also to release potash. With regular use it can prevent attacks of the club root disease (*Plasmodiophora brassicae*), which can only thrive on acid land. It is also useful in checking attacks of some pests.

In the garden it should always be used for all the members of the cabbage family and for the pea and bean family. It is not so important for root crops nor for potatoes. In fact, it predisposes potato tubers to attacks of potato scab. In the flower garden there are a number of plants such as the rhododendron and azalea which dislike applications of lime.

WHAT DOES pH MEAN?

Scientists have agreed to express the degree of acidity of soil by what to the outsider seems a mysterious notation – the pH scale. pH7 represents the neutral point. Figures less than 7 indicate the degree of acidity, while figures more than 7 indicate the degree of alkalinity; pH4, for example, is much more acid than pH6. In this country the range usually runs between pH8.5 and pH4.5. Most crops do best in soils which range from pH6 to pH7.5. Any gardener can test accurately for acidity by using the right indicating fluid, and the colour to which that fluid turns on contact with soil shows the acidity or alkalinity present. A BHD Soil Indicator, which may be purchased from the horticulturist chemist, is a useful set for the home gardener or allotment-holder.

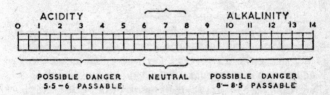

The pH scale.

Fellows of The Good Gardeners' Association can have their soil tested free by the association's chemist.

Different crops will, of course, grow under varying acid or alkaline conditions. Peas and beans, for instance, do best when the pH is between 6 and 7; potatoes, on the other hand, will grow very well indeed when the pH is from 4.7 to 5.7. The Soilometer sets with the indicator fluid usually have with them a little leaflet advising the amount of lime that ought to be used for the varying crops, depending upon the colour of the liquid produced after contact with the soil.

To raise land which you have found to be pH4 to pH7, would require about 2.5 tonnes of lime per hectare (4 tons or 4.48 US tons per acre) in the case of sands, and in the case of clays, 7.5 tonnes/hectare (12 tons or 13.44 US tons per acre). To raise pH6 to pH7 would take about 1 tonne of lime per hectare (1.5 tons or 1.68 US tons per acre) in the case of sands and about 1.9 tonnes/hectare (3 tons or 3.36 US tons per acre) in the case of clays.

It must be remembered, however, that excessive lime can cause trouble. Yellowing of plant leaves may be brought about by this. If so, it is then called lime-induced chlorosis. This is very common in fruit trees grown on the chalky land in some parts of Kent, for example, and is due to the fact that the trees are unable to take up the iron they need because of the excess lime present. It is essential to consider the application of lime with direct reference to the crop that is to be grown.

LOSS OF LIME

There is always a certain loss of lime each year, more as a rule from the sandy soils than from the clays. For instance, in the west, where the rainfall is heavy, the annual loss of lime through drainage may be 750 kg/hectare (6 cwt or 6.72 US cwt per acre). The use of ammonium sulphate neutralizes lime and for every 50 kg (1 cwt) of this chemical applied, 25.4 kg (0.5 cwt) of lime disappears.

During the last twelve years I have discovered that once the soil in the vegetable garden has been neutralized with lime, the composted vegetable waste retains sufficient lime to keep the land in the right condition.

WHAT FORM OF LIME SHALL I USE?

Limes can be classified into three groups: the carbonate, the oxide, and the hydroxide. The carbonate (calcium carbonate) is the form in which lime is found in Nature, generally as chalk or limestone; if these substances are ground to powder they are sold as ground chalk or ground limestone. The oxide (calcium oxide) is really the burnt chalk or burnt limestone. During the burning the carbon dioxide gas is expelled and pure calcium oxide is left. If this burnt lime is allowed to slake, that is, to take up moisture either from the atmosphere or from water purposely given, it turns into the hydroxide (calcium hydroxide) and is then usually sold as hydrated lime.

Calcium oxide is commonly known as quicklime, burnt lime, Buxton lime, lump lime, cob lime, shell lime, and so on. When ground to a fine powder it is known as ground lime. (Please note the great difference between ground lime and ground limestone.) Quicklime (calcium oxide) is the most concentrated form of lime that can be applied. As it quickly absorbs moisture and becomes converted into slaked lime it is very difficult to store. When it can be obtained and applied immediately, it undoubtedly gives very good results.

Hydrated lime (calcium hydroxide) is the commonest lime sold to gardeners and allotment-holders. It is usually higher in price than quicklime, but it is clean, safe and very effective to use. It can easily be stored and, unlike ground lime, has no scorching effects. It undoubtedly has a quicker effect than carbonate of lime.

Limes are usually compared on the basis of their calcium oxide content. Thus if it is advisable to apply 1 tonne (1 ton or 1.12 US tons) of quicklime the pH indicator always refers to the amount of quicklime to be used) it will be necessary to give 1.25 tonne (1.25 tons or 1.4 US tons) of hydrated lime, or if ground limestone is used, 1.75 tonne (1.75 tons or 1.96 US tons). Roughly speaking, therefore, the gardener may say that where quicklime or ground lime is advised, he has to use half as much again if he prefers hydrated lime, and almost twice as much if he wishes to apply ground chalk or ground limestone. This, of course, should affect the price he is able and willing to pay for it.

The lime should be applied on the surface of the ground and never dug in. It should not be mixed with acid artificial manures nor with farmyard manure. It is, however, useful sprinkled on the compost heap – only, however, as advised in Chapter 7.

It has been suggested that if 240 g ground lime were applied per square metre (7 oz per sq yd) once every three years to the cabbage family 'break' in a rotation, then the soil should keep in a fairly good condition. In a four-course rotation, 136 or 170 g/m² (4 or 5 oz per sq yd) is usually given to the cabbage break and 68 or 100 g/m² (2 or 3 oz per sq yd to the peas and beans break, while the root crop and the potato crop would not receive lime.

SPECIAL NOTES AND THOUGHTS ON LIME

1 Unless you have a soil that is rich in chalk, you will have to use lime, except of course where you are growing plants like rhododendrons, azaleas and heathers, which will thrive on acid soils.

2 Get hold of a BDH Soil Indicating outfit and make soil tests in your vegetable garden, or have the soil tested by the GGA soil chemist.

3 Be careful about buying 'waste limes' from 'soap and paper' works, and from acetylene gas works. The former will contain soda and caustic lime and the latter compounds of sulphur and phosphorus.

4 On the other hand, powdered chalk from water-softening plants of electricity-generating stations may be 98 per cent calcium carbonate, and so is good. Unfortunately it is usually only available as a wet sludge.

5 A similar sludge is obtainable in East Anglia and other parts from sugar beet factories. It contains about 30 per cent calcium carbonate. Use it at twice the quantity of dry calcium carbonate.

6 Powdered chalk from the manufacture of whiting is 95 per cent calcium carbonate and can be obtained as a fine powder. Use it as if it were calcium carbonate.

7 Always use lime on the soil as a top dressing – it washes through very quickly.

8 Never over-lime as you may cause magnesium deficiency, and iron deficiency, seen in the yellowing of the leaves of the plants and trees.

This beautiful herbaceous border at Arkley Manor has never been dug. After planting, compost was spread over the soil 1 in deep.

Above Four wooden bins at the demonstration gardens of the Good Gardeners' Association at Arkley. Each bin makes 10 tons of compost. The one on the left has just been filled. The six-month-old compost in the next bin is being used.

Opposite above Compost or sedge peat when used for shrubs gives marvellous fibrous rooting.

Opposite below When cabbages are pulled up to go on the compost heap, there is often a stone or two on the roots. The 6-in layers of vegetable waste are still visible at the end of the six-month period.

Above Compost in the heap ready to use.

Opposite The late Miss Maye Bruce, the inventor of the QR activator, pours the potion into a small compost heap made of wire and corrugated paper. Such a bin does not make successful compost. It is too small and too weak.

Above This heather garden has never been dug or hoed. Compost spread 1 in deep has resulted in healthy plants and healthy soil.

Opposite The author inspects a typical compost-grown row of runner beans. The crops are extraordinarily heavy.

Above The brown powder which is compost.

Below Compost-grown roses. No digging. No hoeing. No weeds.

CHAPTER 10

The Advantage of No-Digging

It isn't quite true to say that I have become a 100 per cent no-digger and no-hoer. Occasional hoeing is done in the vegetable garden at Arkley Manor, but there are rose beds and shrub borders, on the other hand, that have never been forked or hoed for fourteen years. The iris borders have been down eight years without being forked or hoed and they are quite weedless. The heather borders have been planted longer than this and they are equally free. The vegetable garden however has not been dug for fourteen years.

The method here is simply to cover the soil with medium-grade sedge peat or powdery compost. This not only smothers the weeds and prevents them from growing, but keeps the moisture in the soil for the plants to use, while the worms take in the amount of organic matter they need, drawing this deep down (very often) and actually doing the digging for the garden.

Organic manure was never intended to be used to feed plants direct. Organic manuring is adequate only when plants can obtain all their requirements through the biological agencies in the soil without artificial aids. The report of the Ministry of Agriculture, Fisheries and Food's experimental station at Efford, Hampshire, makes quite clear, for instance, that the strawberry, in particular, is an efficient scavenger and that money is therefore best spent on bulky organic manures and not on balanced chemical fertilizers.

Permanent crops should not be disturbed at all. Let the roots of these plants come right up to the surface of the ground and make good use of the rich soil they find there. Never cultivate among the blackcurrants, raspberries, gooseberries or redcurrants. Never cultivate under roses, shrubs or trees nor among herbaceous plants. The soft fruit can be mulched with straw 0.3 m (1 ft) deep. In the flower garden, mulch the shrubs, roses, heathers, irises, primulas and all the herbaceous plants with medium-grade sedge peat 25 mm (1 in deep and leave them alone.

It is true that it may be necessary to replant the herbaceous border every six or seven years and then, of course, the powdery compost or sedge peat will have to be disturbed, but what a tremendous amount of time and labour has been saved meanwhile and how very happy the plants have been in consequence.

In the vegetable garden it is sufficient to cultivate very shallowly. Those who need to save time will probably use a rotivator and so will churn up the top 50 mm (2 in) of soil in the spring. It is never a good plan in the case of heavy soils to do rotivation in the autumn, lest the soil sets down hard. The well-rotted compost that has been made by the keen gardener should be applied all over the surface of the ground at the rate of one 10 l bucketful per square metre (2 gallons per sq yd) in the autumn and any rotivation or raking-in should be done in spring.

There is still a fair amount of argument, even among those who believe in composting, as to the exact definition of no-digging. Some say that the soil should not be cultivated at all. Others argue that a light raking or forking over is of value. There is no doubt that as far as soft fruit is concerned, and most hard fruit for that matter, cultivations are quite unnecessary after the first few years, though if grass is grown, this must be kept very short and has to be fed well.

In the autumn of each year, the non-digger places a 25 mm (1 in) layer of compost all over the surface of any vacant ground there may be. He looks upon this thick layer of compost as the equivalent of the winter digging and manuring. In the spring the small seeds are sown in drills which are drawn out in the residue of the compost and are then covered over with more compost. The larger seeds like the beans and peas are set out in the usual way as if they were in the bottom of the drill and then they are covered with a further 50 mm (2 in) of well-rotted compost.

The late George Copley who carried out no-digging experiments stated that in most cases he found that the actual yields were 50 per cent more than they were by ordinary methods, and that the vigour and quality of the plants were unbelievable. He has grown sweet peas, dahlias, and chrysanthemums on this no-digging idea with wonderfully healthy and floriferous results. He reported that diseases and pests were banished under the no-digging system. He quoted the garden of a Mr A. Guest, of Middle Cliffe, near Barnsley, where a heavy infection of club root disappeared as the result of this modern method.

At the Royal Show at York in 1948 there were two vegetable plots side by side. The one had been dug and manured in the ordinary way and the other had been cultivated under the no-digging system. On the plot that was dug and manured in the normal way the broad beans were alive with blackfly, the carrots had been attacked by the carrot fly maggot, the beet leaves were riddled with leaf miners, and the brassicas badly eaten by the caterpillars of the cabbage white butterfly. There were no pests or diseases, however, to be seen on the undug plot and the seeing was certainly believing. No one was able to explain the phenomenon away – it was a fact and only time and careful research will be able to account for the results.

It seems that an undisturbed soil has some factor in it that may make the difference and, as was suggested above, the very fact that an efficient mulch is applied on the surface of the ground does enable the roots to travel about in the top 25 mm (1 in) or so of soil without being disturbed at all. The no-digging pundits argue that in Nature no attempt is made to bury its waste save by the natural process of gravity and by the functions of the worms. Nature never uses a fire to reduce bulky material to ashes, but waits for the natural decomposition in order to prevent waste.

The no-diggers often use old sawdust as a top dressing in addition to good compost, but most aver that the former should always be preceded by compost for a year or two in order that the population of the particular species of bacteria responsible for the composition of cellulose may be increased. Once this has happened it is possible to put 150 mm (6 in) of really old sawdust all over the surface of the ground with deleterious results.

In firm soils plants develop their natural rooting systems and provide not only themselves with the necessary anchorage roots, but also all the fibrous feeding roots which can come right to the surface for the absorption of the necessary diluted foods. The organic waste which is similar in construction to a sponge is never compressed by top application as it is when it is dug up under the first spit of soil and so is able to absorb far more water. Digging may easily block up the earthworm tunnels which, as we have seen, prove to be the best drainage system possible.

Do not think that in a firm soil there is lack of oxygen, for this will travel happily down the tunnels of the worms. It is suggested that certain species of fungi are very valuable in building up the health of plants, and in fact some aver that they are even more valuable than many species of bacteria, in ensuring immunity from attacks of pests and diseases. The no-diggers believe that the top application of compost makes exactly the right conditions for the beneficial fungi in the soil, and this is one of the reasons why the plants on the non-dug plot at the Royal Show were pest and disease free.

Of course, the regular top dressings of compost do help with weed control, and properly made compost is weed free because the seeds have been killed in the heat engendered in the compost heap. Digging invariably brings weed seeds to the surface, for left alone they will remain dormant if buried for very many years.

A field I know was dug up in 1950 and produced literally millions of that wretched weed called persicaria. Even though the oldest inhabitant did not remember seeing that weed before in the field, the seeds must have laid buried

Charlock.

there for a generation and when the soil was inverted they came to life and became a perfect plague.

THE COMPLETE NON-DIGGER'S BELIEF

Perhaps I could summarize the views of non-diggers as follows:

1 Earthworms do the tunnelling or spading better than the plough.
2 Weeds if properly controlled help with soil aeration also.
3 Properly composted material put on the surface of the ground will ensure better flavoured vegetables.
4 This method will ensure a greater freedom from pests and diseases.
5 The important thing is to keep the 'workers' in the ground happy, and that if the soil can be kept unfrozen the

denizens in the earth will be able to function throughout the winter.

6 The non-digger aims for quality, rather than size, though he may get size as well.

7 Most agree that they are not out to prove that orthodox principles are wrong, but that they have found a better method.

There are always readers who are keen on no-digging methods, whether on a small or large scale. The general scheme is therefore included in this chapter.

NO-DIGGING AND SEDGE PEAT

For many years the author carried out simple experiments with no-digging, and he discovered that one of the problems that faced the man or woman who sets out to work out these theories concerned the actual making of the compost. In the normal way a good garden owner will use compost at the rate of, say, 75 tonnes/hectare (30 tons or 33.6 US tons per acre), whereas if he decides not to dig and to apply the organic matter *completely* as a surface dressing, he really needs about 250 tonnes/hectare (100 tons or 112 US tons per acre).

Most of us are able to make compost. We rot down all the vegetable waste we have. Some are able to buy in straw or even to get waste material from greengrocers' shops. But despite everything that is done in the normal small garden, it seems there is never enough home compost to go round. It was with these thoughts in mind that our Trials Officer decided to see whether it would be possible to use some substance which could adequately take the place of compost, sedge peat being chosen as ideal.

This proved better than any other type of peat, because each tonne contained one-tenth of immediately available humus and had an acidity of pH5.5 – almost neutral. Other types of peat invariably have an acidity as high as pH3.9. They contain about 20 kg of immediately available humus per tonne (45 lb per ton).

For the sake of comparison two small plots were chosen where it was known that the soil was poor, and yet constant right the way through the piece. The crops chosen were peas, French beans, spinach, lettuce, carrots and beetroot – two leaf crops, two members of the pulse family and two root crops. In the case of each plot, the exact amount of sedge peat was weighed out so as to give a coverage of about 50 mm (2 in) all over the ground. Then in the dug plot the peat was forked in, and in the undug plot it was just left on the surface of the ground.

The sowing of the seed was carried out in the normal way in the forked-over area and in the undug plot the drills were drawn out in the peat, that is to say a small narrow hand hoe was used to produce what may be called a 'slit' in the sedge peat so that the seeds could be laid on the surface of the ground. Another method used is to put the seeds on the ground and then put the peat on top afterwards. The peat was applied at the same time in these two cases, however, so as to make the two plots comparable.

It was decided not to use any organic fertilizer in a dry form, but to apply an organic liquid fertilizer liberally once a fortnight when the plants were through the ground. This, when diluted, was given through the fine rose of a can right the way down the rows of plants, equal quantities being applied to both plots on the same day and at the same time. The peas were supported in the ordinary way, and in the case of the French beans, the rows of the undug plot had to be given some support, because the foliage grew so luxuriantly.

As regards results, one can truthfully say over three years that there was very little difference in the results between the dug and the undug plots. The beetroot and the peas did better on the undug area in all the years, but there was little to choose from in the case of the spinach, the French beans and the carrots. I preferred the lettuces on the undug plot and especially so in the second and third years. From the third year onwards the vegetables on the undug plot proved far more delicious and were free from pests and diseases.

For all compost gardeners and particularly for those who are old or infirm and do not find it easy to fork or dig, this method of growing vegetables is particularly suitable. The use of sedge peat certainly gives pleasure, for it smothers weeds, creates a mulch and is beautifully clean to handle. It is obviously ideal in the case of those who do not find it easy to make compost.

This scheme has been adopted under glass, and it seems an excellent plan for tomatoes, especially in heavy soil.

There is no need to do any digging, forking or hoeing in the flower garden. The roses, shrubs, herbaceous plants can be planted and the sedge peat put all over the soil at least 25 mm (1 in) deep to prevent the growth of the annual weeds.

At the Arkley Manor gardens there are large beds of heathers, ferns, roses, irises, primulas, kniphofias, phloxes, etc, which are growing according to the sedge-peat mulching method since planting in 1960. The beds look attractive and the plants couldn't be better.

NO HOEING AND THE REASON WHY

Numbers of people are adopting no-digging and no-hoeing methods, or what may be called perhaps the minimum-digging methods, and they're getting better results from them than from the old-fashioned ways of digging the soil. The main criticism (and it's a baseless one) is that those who adopt this method are bound to find that the soil is unsatisfactory, because it's never been aerated, and they say you can't have a good soil and sow seed without digging or ploughing. What happens, in fact, is that by putting the compost on top of the ground the earthworm population greatly increases, attracted by the compost. Regular burrowing, by day and night, with a huge force of worms, provides better aeration than could possibly be obtained by digging or forking.

I was making an estimate count on my land some time

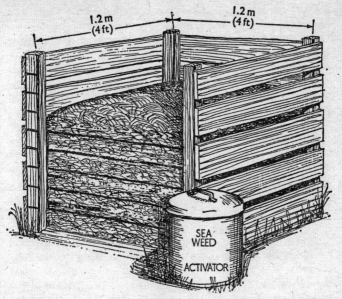

A correct square wooden bin with air spaces between the planks. The marks on the post indicate the 150 mm (6 in) layers.

ago, and I've got roughly about 30 million worms per hectare (12 million worms per acre) at the moment; and what the build-up's going to be in the next three or four years I just don't know, because one worm can produce 600 worms in a year, if the soil is right.

WEED POPULATION

In a normal 0.2 hectare (0.5 acre) garden 174 million weed seeds can be buried in the top 100 (4 in) of soil. Even if only 15 per cent of these weeds are viable – that is to say, are alive – it isn't surprising that thousands of weeds flourish in our gardens. The British National Vegetable Research Station have been making counts of weed seeds in various market gardens, and they've issued some very surprising reports. They found that in some gardens there could be 75 kg (1.5 cwt) of one type of weed seed alone. Remember, too, that if a gardener leaves one weed seed, then it may grow into a plant that may produce hundreds more weed seeds that very year.

I've discovered that it is possible to have 430 weed seeds per square metre (40 per sq ft) and thus you can easily have 80 million weed seeds per 0.2 hectare (0.5 acre). In the case of the weed called the annual meadowgrass, 90 weeds per square metre (8 per sq ft) will produce 180 million weed seeds per 0.2 hectare (0.5 acre) the next year. In fact, in one garden I knew 50 weed plants per square metre (5 per sq ft) produced at the end of the year 122 million weed seeds per 0.2 hectare (0.5 acre).

Every time you hoe or cultivate the surface of the ground, you're bound to bring up weed seeds, and these are seeds which will germinate and produce weed plants. Nobody wants weeds, they want the plants they are growing. The method adopted by the Fellows of The Good Gardeners' Association (and there are branches in many different parts of the world) is *no hoeing at all*; no weed seeds are brought to the surface of the ground, and there aren't any weeds.

To get this kind of result, perfect compost must be made. A brown-black, powdery compost, free from disease, pests and live weed seeds, must be produced. This can easily be done if an organic activator is used and if the composting is done in a correct square wooden bin. Thus a lovely brown powder, free from disease, pests and weed seeds, is produced and this goes on the ground as mulch.

CHAPTER 11

The Job Done by Weeds

We all of us want to keep down weeds, because they compete with the plants that we are trying to grow, but weeds have something to say to us, if we look at them intelligently. For instance, some weeds point out that the soil is acid: sorrel, for instance, corn marigold, spurrey, the scentless mayweed, bracken, sow thistle, coltsfoot, nettles, and the wild pansy. Other weeds show that the land is limey, or, as some people put it, chalk loving, and here you find the wild mignonette, white mustard, the musk thistle, the wild carrot, and henbane.

Sometimes, as an adviser, the author walks over a piece of land, and finds horsetail, fitches, mare's tail, redshank, and the cotton grasses, and then he immediately knows that the ground is waterlogged, for these are the weeds that are called moisture loving. A walk over another field, and immediately one discovers that it's a heavy soil, because there will be found the creeping buttercup, the dandelion, the creeping bent, and the coltsfoot, whereas if these weeds were absent and one found instead the small nettle, and perhaps couch, or twitch, then the indication would be that the soil was light. The appearance of lots of vetches, together with the kidney vetch, speaks of nitrogen deficiency. Though bracken may indicate that the soil is lacking in potash, curiously enough it has the power of obtaining potash out of potash-deficient land, and so when put on to the compost heap gives potash liberally.

The nettle, on the other hand, somehow manages to extract lime from an acid soil, while lupins have the power of gathering zinc. It is not really as mysterious as it seems, for what happens is that the plants search out over a far wider area than the chemist does in any soil sampling which he manages to make. It can therefore be said quite honestly that weeds can be very useful when put on the compost heap with an activator.

Stinging nettle.

A weed, of course, must be described as a plant that is growing where it is not wanted. Actually, however, we have come to look upon certain plants as weeds – the dandelions, the shepherd's purse, nettles, groundsel, and so on, and we regard them as pestilential nuisances. We only wish that we were in the Garden of Eden where there were no weeds and only the right plants grew in the right places. It is, therefore, interesting to realize that even weeds have their functions, and a study of them reveals some interesting facts.

In the first place, weeds have been described as silent miners. Their roots go down and help to break up the soil. Take that rather pretty weed known as chicory or suckery. It has a lovely blue flower something like that of a dandelion in shape. Its rosette of leaves near the ground is dandelion-like too. It doesn't rob the plants round about as so many people think. It's roots are surprisingly long and go down and down, seeking for water supplies that other plants cannot possibly get. This tunnelling is useful, and then as the roots start to swell it moves the soil and opens it up helping to increase the warmth. Its silent mining is doing a great deal of good by letting in the air and helping in the drainage.

Dandelion.

The dandelion is another deep rooter. The yarrow is different in that it has roots that creep just below the ground as well as large numbers of slender tap roots lower down in the soil. I like it as a weed because it produces lots of vegetable organic matter for digging-in or composting. The gardener who is a digger (and I am not) digs to the same depth each year, ie the depth of his spade. The useful deep-rooting weeds, however, take no notice of the depth of cultivation – they go well beyond the spade's depth, boring into the subsoil, helping to break it up, and cultivating and aerating it.

There is no doubt that they are valuable agents for promoting water absorption and warmth. They help to divide and push the soil about. They help to let water in, and they help also to leave fibre behind which improves the soil. Naturally certain of the cultivated plants have a similar effect, as readers will no doubt have said to themselves as they were reading the previous paragraph.

WEEDS AS INDICATORS

As I have tried to explain in the introduction to this chapter, weeds have their value also in indicating the condition of soils. Some weeds have a particular ability to accumulate some definite mineral substance or substances which may be deficient or lacking in the very soil in which they grow. The weeds growing in ground that may mean starvation to a normal crop develop a striking efficiency for gathering in chemicals in a famine environment. How certain weeds are able to do this is not understood, but it may be that because they are the products of their environment they are able to fix certain mineral elements from the air, very much in the same way that peas and beans and other members of the leguminose family are able to fix nitrogen from the air.

Weeds natural to a particular soil therefore have a definite function, because in the normal way they will die and rot, and the substances which they have abstracted will be delivered up to the soil, thus enriching it. This is the reason it is so important to compost weeds instead of burning them. Weeds can therefore be said to have a practical value in that they provide the traces of mineral foods which plants definitely require.

On chalky soils, the weed known as henbane is very common. Though the soil may be rich in lime the henbane contains very little lime at all, but is rich in phosphorus. Rich, well-cultivated gardens usually abound in groundsel, chickweed and annual nettles. Their function seems to

Groundsel.

be to gather together silicic acid, a 'food' which tends to become deficient in gardens that have been worked for many years.

Sorrel always grows in acid ground, but, in fact, is always rich in calcium (lime). There are other plants rich in lime which are given the name of calcium-efficiency plants. These include the plantain, the spurrey and hearts-ease, while on acid garden lawns will be found dandelions and daisies in addition to the plantains, all of which indicate acidity and yet contain calcium. The members of the leguminose family, like the vetch, rape and clover, always indicate that the soil is deficient in nitrogen, and yet they themselves will be rich in this plant food as will be found when examining their roots, for there will be seen the white nodules.

There are a number of weeds which indicate stagnant conditions in the soil, bad drainage, excess of soil moisture, and so on. On the surface will be found the mosses and the tussock grass, while in addition such weeds as mare's tails (*Equisetum*), rushes and sedges will flourish, and so, often, will meadowsweet.

Weeds like yarrow and chamomile, containing as they do potash, indicate potassium deficiency in the soil. The yellow dock (*Rumex crispus*) is rich in iron, so is chicory. The wild marigold contains sulphur; and the wild strawberry, calcium and phosphorus.

Sufficient has been said to show how weeds have their part to play and must not just be regarded as competitors of the chosen cultivated plants. The successful cultivator will do well to learn weed recognition, for by them it will be possible to judge the conditions of the soil to be dealt with.

CHAPTER 12

Organic Methods of Fruit Growing

When The Horticultural Training Centre moved from the Swanley district in Kent to Thaxted in Essex, and became known as The Thaxted Horticultural College, we determined to start all our fruit-growing activities on a new basis. Much had been learned in Kent, and there was no need to repeat mistakes. The main orchard was therefore planted with the intention of grassing it down in three years' time, and this was done with the seed mixture which is detailed later on in the chapter. A fruit cage was established in the walled-in garden, on the straw mulch system, which proved very profitable over eleven years. Each season a little straw was added as a top dressing and as a result really heavy crops were gathered, despite the fact that there had been no hoeing, no forking, and no digging.

In the main fields, the soft fruit has been grown in a similar manner, with the exception of the blackcurrants, which after three years were changed over to the grassing system, a system very similar to that advised and adopted for the apples and pears, and which merely means that the grass around the bushes is cut eight or nine times a year, and that quite heavy doses of a fish manure are applied in the spring and in the early autumn. We have therefore been able to show that good fruit can be grown on the no-digging method, and furthermore in years like 1973 when the early part of the summer was extremely dry, we had heavy crops of blackcurrants, of good size and flavour, when very few growers were able to pick them satisfactorily. The mulchings had done their work, the roots had been able to develop at will, without being disturbed at all, and this meant, and has meant since, that the bushes are happy and so give fruit of excellent flavour in abundance.

It was my privilege to have as a friend the late Mr B. S. Furneaux, the famous soil surveyor. No one has done more for the horticulturalist, as regards the use of soil, than this expert. With his well-known soil auger and his extensive knowledge, he is able today to assess the value of almost any soil in, so it seems, a few moments. There is no one in whom I have greater confidence when it comes to having to make a decision as to what crops to grow on any particular piece of land.

It is therefore not surprising that I am taking the liberty of quoting freely from conversations I have had with him as well as from his writings. It can be said that good soils and bad may easily lie side by side, and this is why it is extremely important to plan carefully when planting a fruit crop which is designed to occupy land for a long period.

Almost all kinds of fruit prefer a well-drained soil of adequate depth but whereas it can be said that some definitely demand it, others will grow and crop satisfactorily on some soils that may fall short of this ideal. Cherries are very fastidious – they prefer a medium-textured soil and perfect drainage. There must be a good depth too, and what is called a brick earth usually complies with these specifications. There are some varieties that are more easily accommodating and especially Early Rivers, which curiously enough will not only put up with some poor drainage but will also grow on a heavier type of land. Noir de Guben is similar in this respect and it is pleasant to be able to record that these two varieties will pollinate one another. Sour cherries do well on sandier soils and on a medium loam but they also insist on good drainage.

In the case of dessert apples, it is necessary to have a soil of a depth of, say 0.5 to 0.6 m (18 to 24 in). Good fruit may be grown on all kinds of soils fortunately. Cooking apples are not so fastidious about good drainage as are desserts – but the deeper the soil the better. Pears grow quite well on a great variety of land and they will put up with poor drainage providing the soil moisture is not charged with lime. Plums do best on a deep well-drained soil, but if they are well fed they are tolerant of a degree of poor drainage. They may be put into the same class as cooking apples on the whole.

Drainage is very important for gooseberries and especially so for that delicious dessert variety Leveller. The good cooking variety Careless is far more tolerant of wet soil. In the case of blackcurrants it is surprising the amount of bad drainage that I have seen them put up with. As against this, most varieties prefer the medium or lighter types of soil, though Westwick Choice and Wellington XXX do quite well on heavy land. Raspberries like a deep soil and a good drainage. Loganberries and blackberries must be planted on soil which doesn't encourage autumn growth, because of frost damage to the canes, and strawberries like a soil which retains moisture well though they, like blackcurrants, can put up with a considerable degree of poor drainage.

Man's management is the overriding factor in almost any soil, and to talk in a general way of a good soil or a bad soil, is often very misleading. Every soil has a range of performance; properly managed, every soil can be cropped. Of course it pays best to plant on the most promising soils, but perfection even in soil is seldom obtainable. Drainage isn't always a question of laying down agricultural pipes; it may be good enough to clean out the ditches or to fork up a hard pan which may be lying low down – this can sometimes be done during bastard trenching.

Soil is more than a mixture of chemical compounds – it must contain organic matter. Humus need not be endowed with mystical properties but it must have conferred upon it a very important place in the maintenance of soil fertility. It is for this reason that we believe in the grassing down of soil below fruit trees and mowing the grass regularly so as to return great quantities of organic matter to the ground each year. This will encourage the multiplication not only of the earthworms but also of the beneficial soil fungi and, as we have already said, it is these living organisms which ensure the free ingress of water and air and also the free movement of roots.

115

Air, water and roots share the complicated communications system of the soil. They can be likened to our own network of roadways – earthworms, which penetrate a number of feet into the ground, construct the arterial roads. The soil itself is divided by a number of fissures which correspond to the secondary roads; the individual blocks of soil are traversed by minute holes by the decay of roots; these are the country lanes. It is upon the development of this system, which in itself is dependent largely upon the amount of decaying matter, that the full efficiency of the soil depends. Earthworms, too, with their penetrating shafts, tend to deepen the soil, pulling down organic matter even well into the unweathered geological material beneath the soil.

Some years ago, the experts Dr W. S. Rogers and Mr D. W. P. Greenham put forward the view that because there was an increasing tendency to site fruit orchards on hillsides in order to avoid frost damage, soil erosion had become a more serious problem in England. They said that the maintenance of organic matter was the only practicable method of ensuring those soil conditions necessary for healthy plant growth and as such, this maintenance was of paramount importance. Organic matter was the regulator of the water supply. It acted as a kind of chemical sponge, a reservoir of nutrients more or less ready balanced. It affected the biological condition of the soil. It affected the earthworm population. It played a most important part in the maintenance of soil structure, for it supplied the energy material necessary for the activity of the micro-organisms which produced crumb structure in soils. Further, the presence of organic matter helped to prevent soil erosion.

SOWING THE COVER CROP

Some gardeners like to plant the trees in cultivated land and then to sow a grass mixture five years later in April or September. A typical seeds mixture is ten parts by weight of perennial ryegrass, one part Kentish wild white clover, one part of white clover S.100, four parts of broad red clover, two parts of sheep's fescue and two parts of smooth-stalked meadow grass. This mixture should be sown at about 34 g/m^2 (1 oz per sq yd). When the plants have grown, the sward must be kept short by frequent lawn mowings and once a year, probably January, dried blood should be applied at 100 g/m^2 (3 oz per sq yd). This recommendation particularly applies to apple orchards.

Another simpler method is to use a very fine type of grass known as chewings fescue. This grass when left alone lies flat on the ground like so much hair and acts as a natural mulch. This can be sown at as little as 17 g/m^2 (0.5 oz per sq yd), and being a fine-bladed grass will grow on the most dry soil. Once again the grass must be cut regularly and the mowings allowed to be returned to the soil. They must not be carted away in the normal box such as one uses for lawn mowings. Try and mow the grass about once a fortnight in the spring and summer when it is growing well.

It can be said that apples and particularly dessert varieties growing on grass have much better colour and better flavour, and it does seem that this 'organic method' of growing apples is not only considerably cheaper as regards management expenses, but in addition the results are far better. At the Arkley Horticultural Centre we are carrying on with experiments on plums to see whether we can manage them in the same way. The idea certainly holds good with pears and with sweet cherries.

THE STRAW MULCHING METHOD

With soft fruits the scheme is to use quantities of straw as a mulch. Naturally the cleaner the straw used the better and, generally speaking, wheat straw is better than barley or oats. The plan is to put the straw down to a depth of 150 mm (6 in) during the month of November. The straw mulch must go all over the ground, not just around the bushes, with the result that the weeds are smothered and the bulk of the moisture is retained in the soil. The surface-feeding roots are able to come right up to the top of the land and are never disturbed at all, for no digging or forking is done. The idea will certainly please the no-digging brigade.

No attempt is made to remove the straw at all, but, of course, no smoking is ever allowed in the soft fruit garden. Gradually the straw rots down and much of it is pulled into the ground by the worms. If any weeds do poke their heads through the straw mesh, they have to be pulled up by hand. Few of them, however, do manage to make an appearance. Each year a further dressing of straw is put into position if necessary during the month of November and, as a result, all expenses concerned with cultivation are saved. The only exception to the rule is perhaps the gooseberry which must be mulched later, say, in the month of March, because if the straw is put on too early, there is a tendency for the fruit to develop later. Those who desire earliness, and most do, should therefore avoid the earlier period of mulching for gooseberries which, however, is suitable for the black-currants, redcurrants, strawberries and the cane fruits like blackberries and loganberries.

In addition to the straw, an organic fertilizer will be used each season during the month of November. This should be one like Meat and Bone meal, applied at the rate of 100 g/m² (3 oz per sq yd) plus wood ashes, at 270 g/m² (8 oz per sq yd). Alternatively, potassium sulphate at, say, 68 g/m² (2 oz per sq yd) could replace the wood ashes, but is not so effective. There are potassium sulphate supplies on the market today which are organic in origin. In very dry weather, and this is the exception rather than the rule, it might be necessary to put the overhead sprinkler in position in June and to give a very thorough soaking. This, however, as I have said, would be an exception, applying principally to the eastern counties of England.

CHAPTER 13

Health from the Soil

How long is it going to be before the housewives and
mothers of the world revolt against the adulteration of the
food they buy day by day and week by week? This is
where they truly are the controllers, for not only do they
purchase, but they prepare, the food that is *supposed* to
sustain and nourish our bodies. In fact, if Women's Lib is to
mean anything at all, it should lead to something practical
in the realm of human health.

Do people these days expect to be well or do they know
that they will have to visit the doctor? Are they sure that
they will have to take some vitamin injections or certain
pills? What about constipation? What about arthritis?
How many friends are suffering from fashionable diverti-
culitis?

Man may be flying into space and landing on the moon,
but millions a year are spent on patent medicines to try
and solve his health problems. The British alone swallow
ten million aspirins a day – and this is only one of the many
pills men and women buy. In about sixty years of medical
progress, the number of patients in British hospitals has
quadrupled, while the population of the asylums has
doubled. What is needed, surely, is positive health – so that
doctors can start to think about disease prevention rather
than cure.

Surely our bodies should be constantly and regularly
renewed and this perfect renewal must be through the daily
flow of nutrient materials from the soil *through* the plants
and the birds and animals. It is quite obvious that we must
look in the first place to Mother Earth and to see that she,
from whose womb perfect food can come, is properly fed
and looked after. Then we must tackle the food-processing
procedures. It must be remembered that there is hardly

one item of popular food that is not subjected to a practice at one stage or another which is inimical to health or what I call 'wholeness'.

Remember we are sold what can be called synthetic medicated bread. The wheat grain is crushed with steel rollers which remove from the flour the wheat-germ oil containing Vitamin E. Then, the natural calcium is extracted and chalk added. Chemicals are used to bleach the flour, most of the nutritive part used to feed the pigs and poultry and the extremely important wheat germ is removed and sold to the public at hundreds of pounds per ton.

There is another very serious problem that we face today in respect of food. Farmers are using strong harmful hormone sprays in greater and greater quantities and these are used to saturate the cereal crops in order to control weeds. Unfortunately, they are not always used with care and again and again these sprays drift on to market gardens and fruit farms and severely damage or poison the crops.

I am often called in by horticulturalists all over the country (and abroad as well) who have had their valuable fruits and vegetables ruined by spray drift. Sometimes crops are literally wiped out. The weed-killers and so-called insecticides float in the breeze and land where they will, killing predator friends as well as enemies. The idiotic thing is that in many cases the pests become immune to the sprays and the weeds also, and so new and more virulent hormones and insecticides have to be invented.

Look at the trouble that DDT has caused – no wonder it has been banned in several countries. You have only to use a DDT fly spray for it to pass into our bodies where it builds up cumulatively in our fat awaiting the day when we have some illness. The moment we begin to live on our fat the poison is released and we are in for serious trouble. If our land and our world is to survive we must stop believing that we can murder any creature in the realms of nature with impunity.

I ask all women to take a stand against battery eggs, heat-treated milk, drugged bullocks, chemically boosted bacon, battery calves, and so on. The Good Gardeners' Association was formed so that members could be taught how to grow their own food without the use of any chemicals at all.

There are two slogans that Fellows of The Good Gardeners' Association believe. They are: 'Only a plant really knows what another plant wants. Everything that has lived can live again in another plant.'

Having understood these slogans, ask yourself whether you are one of those who think that a poor soil will produce a small crop and a rich soil will yield a large crop. Do you always think in quantities, without any thought as to quality of the crop? Are you always wanting to produce a lot and not wanting to produce some food, some flower that really is better than anybody else's in its content?

You see, few people realize that there is any connection between the kind of soil their food is grown on and their good, bad, or poor health. People believe that carrots are carrots, and cabbages are cabbages, and they feel that it doesn't matter whether they are grown on what we call depleted soils or whether they are grown on beautifully fertile soils, rich in humus.

Now, the yield per hectare or per acre may be determined by the fertility of the soil, but the quantity of the product, people believe, remains the same, whether it's grown on poor soil or rich soil. This assumption is wrong. Vegetables grown on depleted soils may look exactly like those grown on soils rich in humus, but on chemical analysis, the former may be high in starch and cellulose and low in proteins, while the latter will be low in starches but much higher in the amino-acids, the proteins, the vitamins, the enzymes and the minerals. And since it's known that much of our health is dependent on vitamins and proteins, it's easy to see the connection between the kind of health you and I enjoy – or don't enjoy – and the kind of soil our foods have been grown on.

I have a cousin living in Australia whose health these

last two years has dramatically changed because she is now eating crops grown on the kind of soil which has been treated in the right way. Reports of special trials in soil health done by the Soils Department of the University of Missouri (a pioneer in this type of experimental work) reveal that the amount of carotene, from which Vitamin A comes, varies from as low as 0.5 mg per 100 g from the chemical-fertilizer-treated soil to as high as 31 mg per 100 g of fresh carrots on the compost-only-treated soil. This is a definite pointer.

Experiments carried out showed several times the amount of Vitamins C, B_2, B_1 and A, and even more protein, phosphorus, calcium, potassium and iron in turnip greens that were grown on compost-fed soils than those grown on chemical-fertilizer-fed plots.

All plants require certain nutrients for healthy growth; and of the hundred or more elements known to chemists, some twenty-five of them seem to be essential for the normal plants. There are about twelve more which may not be really necessary to the growth of plants, but which, when present in the plant tissues, are found to be beneficial to the health of the animals that eat those plants. It has been proved that deficiency diseases can be induced in animals by withholding foods containing these elements. Farmers know that, if they are to keep their cows at the peak of health and milk production, they must feed their animals with foods high in vitamins and proteins, such as the wheat germ, the bran and the middlings. These, of course, are removed from the wheat when it's milled for white flour, in order to produce the white bread we humans eat. Thus, animals stay healthy very often, while humans suffer from all kinds of degenerative diseases as the result of eating what are called refined foods, which contain nothing more perhaps than starches and cellulose – containing no more nutrients, perhaps, than in sawdust.

Every bag of chemical fertilizer has three numbers printed on the bag. This represents the percentage of nitrogen, phosphorus and potassium contained therein. These three elements are said to be the macro-nutrients. A score or more of other elements are required for growing plants, but such infinitesimal amounts are needed that we call them the micro-nutrients, or sometimes the trace elements – boron, copper, iron, manganese, molybdenum, zinc, cobalt, iodine, chlorine, sodium, lithium, aluminium, silicon and so on. A few of the elements lie between the macro- and the micro-nutrients whose amounts are therefore intermediate. These are calcium, magnesium, sulphur, carbon, hydrogen and oxygen.

Although nitrogen, hydrogen and oxygen are gases floating in the air above all growing plants, not one particle can be appropriated as food nutrients by the plants. But carbon dioxide and other gas in the air can be taken into the plant leaves through the pores called stomata, and oxygen can be taken in through the stomata for respiration, but not for food. Nitrogen cannot be taken into the plant except through the roots in the form of nitrates and ammonium and organic salts or compounds. Not one of the elements named above as either macro-nutrients or micro-nutrients can be taken into the plant roots as pure elements. They must all be combined into compounds that are soluble in water. The non-metals combine with metals to form salts, or with hydrogen to form acids.

Carbon is a non-metal (being a black substance like coal), but it has the ability to combine with hydrogen and oxygen and practically every other element in a million ways to form organic compounds and organic acids. If the soil is rich in the minerals of both the macro- and micro-nutrients, as in the case of soils fed with compost, these organic acids are taken up by the roots, transported to the leaves, manufactured into amino-acids, proteins, vitamins and enzymes, and, in addition, into starches and sugars. If the soil is poor or depleted of plant nutrients, the stored-up foods in the vegetables and seeds are low in proteins and vitamins and high in starches.

There are other reasons for maintaining a high level of organic matter in the soil. Compost provides food for the worms, and their tunnels through the soil allow the air to circulate to the roots. Their worm casts provide plentiful, soluble nutrients for the plants as well; for both the compost and the worm casts generate organic acids that act on the mineral rock powders, dissolving them and liberating certain trace elements that are needed.

Organic matter acts as a sponge, absorbing the surplus water and holding it for future use in the plant roots. Probably its most important use, however, is as a chelate. Chelate is a Greek word meaning claw, and the molecules of the compost wrap themselves around the insoluble molecules of a mineral rock and can pull them out of the rock like a claw and then proceed to dissolve them. That explains why true compost is far better than a chemical fertilizer for a depleted soil. The fertility may be there in a poor soil, but it's locked up in an insoluble form, waiting to be chelated by the humus.

A poor soil may be stimulated by chemical fertilizers to produce larger yields, but on analysis the foods produced will be mostly starches with very little proteins. But a soil high in organic matter, I find, will produce food crops that are high in proteins, vitamins and minerals, with relatively smaller amounts of fattening starches. There is a farm in Suffolk which is divided into three groups and the cows on the group where the soil is fed with compost only give higher yields of *better milk from less food* than those grown and feeding on the land that has been fed with chemicals only.

Have you ever heard the widespread myth that foods grown on poor chemically treated soils are just as nutritious as those grown on soils enriched by compost? Have you heard, and been told, that foods with all the nutrients removed by processing are just as healthy by the addition of a few chemicals called synthetic vitamins? If you have, then don't believe it any longer. I always say it's exactly like a highway robber taking £100 from you and then returning you a couple of pennies, and saying 'Look! Now I've enriched you!'

Just as the quality of the food you eat comes from the quality of the soil it is grown in, I believe that the kind of health you enjoy comes from the quality of the foods you eat. You are what you eat.

IS ARTIFICIAL REAL?

No one, surely, is going to believe that synthetic rubber is in every respect as good as real rubber made by Nature in the forest. One of the USA universities reported some time ago that 'natural nitrates have something that the artificial ones lack, and there was, in fact, no completely adequate substitute for them in the field of agricultural fertilizers'. They went on to say: 'It is, of course, just as impossible to make artificial nitrates that duplicate natural nitrates, as it is to make synthetic sea-water that contains all the elements of natural sea-water.'

I believe that the crucial test of *real* scientific achievement is whether it recognizes and respects the supremacy of Mother Earth – or ignorantly attempts to substitute the false for the true.

Unfortunately, this substitution goes on not only in the vegetable garden, but also in the grocer's shop. The result is that the sale of anti-deficiency products is really big business, running in the neighbourhood of £2 million annually in the UK alone. This is an indication of how widespread British dietary deficiency is, and how readily money is spent to regain damaged health. (By the way, the sum mentioned does not include the cost of medical treatment, involving special prescriptions. It covers only the sales of packaged vitamin and mineral preparations sold by pharmaceutical chemists.)

Since your body is known to need all of the important natural elements, just how are you to be sure you are obtaining enough of all these factors, major and minor, to be reasonably certain of avoiding the onset of deficiency or degenerative diseases? The only logical way of supplying your body with them is to get them regularly in your customary diet. The body is continually eliminating these elements, if they are present, hence the supply should be just as constantly renewed.

If you are to obtain them from your food, they must be kept available in the soil from which that food is produced. Plants obtain nitrogen and carbon dioxide from the air but they must get the minerals almost exclusively from the soil. Suppose you follow religiously the recommended diet of some authorized authority – won't such a diet protect you against mineral deficiencies? Unfortunately, the answer to this must almost always be 'No'.

Theoretically, such a diet should meet all your bodily requirements. Actually, however, it cannot meet these requirements unless the food has been produced on *completely fertile* soil, ie one that contains all the elements necessary for healthy plants and animals. Thus, a person could consume in excess of the recommended quantities of all the foods listed in such a 'health diet' if they came from deficient soils and still fail by a wide margin to supply all the minerals and vitamins which he requires for maintenance of health. It was this fact that caused the Chairman of The Good Gardeners' Association to coin the phrase 'starving to death on a full stomach'.

There is a very wide variation in the composition of fruits and vegetables and eggs when produced on different soils or in different gardens. Numerous tests have proved this fact, and the difference is often so pronounced that even a layman can notice it. For example, there are those who will not buy vegetables and fruits from chemically fed soil, because they do not have the same feeding value as crops from land where the original mineral content has been replenished year after year by plenty of compost.

Despite all the progress we have made in processing and packaging foods for convenience in transportation, distribution, and home preparation, our knowledge of the actual constituents of various foods is far from complete. This is especially true with regard to the 'minor' or 'trace' elements. While these may be minor in volume, they are often of major importance in their effects upon the health of the man or animal eating them.

The attitude of people today towards these minor elements is much the same as that of the public towards bacterially borne disease in Pasteur's day. Then, the average man in the street seemed to feel that any organism so small that it could not be seen with the naked eye need not be feared. The general opinion now seems to be that elements which are needed by our bodies in such infinitesimally small quantities that the most delicate scientific instruments are required to detect their presence, cannot have much effect upon a person's health.

There is also the feeling on the part of many people that since the human race has for centuries eaten certain staple foods produced on our soils, and maintained a reasonably satisfactory state of health, it can do the same thing today. Those who hold this belief try to ascribe the growing rate of degenerative diseases such as those of the heart, liver, teeth, bones, etc, to causes *other* than the quality of food. There is only one major disease, and that is malnutrition. All ailments and afflictions to which we may become heir are directly traceable to it.

A simple method of assessing food value has been developed by scientists at the University of Texas. Their research shows that free-range eggs are significantly superior to battery-produced eggs (those produced from chickens which are always caged), and that the real value of food differs greatly from what its mere analysis implies.

The nutritional workers begin with the premise that food firstly supplies an animal with energy, and secondly allows it to construct or reconstruct its own body tissues. Unless food has sufficient constructional or trophic value, then obviously it can never allow an animal to realize its full genetic potential even though it may keep the animal alive. But trophic value has proved difficult to measure.

Their method involved measuring the growth of animals fed on a given food. They found that rats fed on free-range eggs reached 350 g in twelve weeks, while those fed on battery eggs reached only 315 g.

The Professor and his colleagues reported in *Proceedings of the National Academy of Science* that brown rice predictably surpasses polished rice; ripe bananas and canned carrots both have low trophic values. Testing food for their trophic value should be obligatory, say the Texas University workers. If it were, many common foods could then be greatly improved.

Plants do perform some near miracles, but they simply can't extract from a soil minerals and vitamins which are no longer there. They will always do their best to grow and produce seed to perpetuate their species, but when the soil is exhausted, the plant cannot enjoy real health.

For many generations crops and animals have been grown on our farm lands and shipped to the cities. With them, at first, went the plant foods which Nature intended they should contain. Each year's crop contained only small quantities of trace elements, but in the one or two centuries that our better soils have been farmed, some of these important elements have been depleted. The plant foods and organic matter shipped to the cities never returned to the soil of the farms and gardens from which they were hauled. Instead they went back to the sea as sewage or landed in inaccessible spots as garbage dumps. Animal bones were never returned to the farms where they were needed to produce health-giving food.

Well over a hundred years ago a famous German chemist, Liebig, analysed a human body. He found calcium, nitrogen, phosphorus and potash, in addition to water. His crude methods showed the same elements in plants and animals and he concluded that so long as these elements were replaced in the soil in generous quantities neither plants nor people would suffer from malnutrition. This opinion from a noted scientist was accepted as gospel truth by scientists and laymen. For over a century few people had the temerity to question his assumptions. His teachings still dominate every classroom where soil chemistry is taught. The value of calcium (usually applied as quicklime, or hydrated lime) has long been known. The present wide use of fertilizers containing nitrogen, phosphorus and potash (NPK) is a direct result of Liebig's work. It may have been possible in the early years, following the propounding of the theory, to get good results with chemical fertilizers alone, for, of course, there was the large original supply of humus in the soil. But the quickened growth induced and introduced by the chemicals also speeds up the rate at which the humus is exhausted; and so today the situation is quite different.

His teachings have been, and still are, responsible for untold sufferings of millions of people whose health has been ruined by deficiencies of minerals in their food. For Liebig's declaration was one of the dangerous 'partial truths'. Calcium and NPK are essential, but they are not the whole story. It now seems clear that, instead of pointing the way down the highway leading to improvement in health of plants, animals and people, he guided us into a dangerously narrow by-path from which we are just beginning to emerge.

Today, modern analytical technique shows that plants contain sodium, potassium, calcium, magnesium, copper, zinc, aluminium, silicon, phosphorus, iron, boron, manganese, barium, strontium, lead, lithium, tin, cobalt, silver, nickel, molybdenum and chromium as well as iodine, sulphur and selenium very often. This listing shows how far we have progressed in the past hundred years in developing methods and equipment for analysing materials.

For years there have been gardeners and farmers who recognized the shortcomings of the NPK formula. Their own observations and experience convinced them that really nutritious food cannot be produced from deficient soils, even though they might be liberally supplied with calcium, nitrogen, phosphorus and potash in a chemical form.

At Arkley Manor, the home of The Good Gardeners' Association, we improve the health-giving qualities of the soil products by adding bacteria-rich powdery compost. Thus we build up the earthworm population, so releasing more minerals from soil particles. We are enthusiastic about the results that are being obtained. We get better quality and flavour in vegetables and fruit crops, and an improvement in the health of the human beings consuming them.

CHAPTER 14

Facts and Theories

Over the years, there have been many theories about the fertility and health of the soil. I have already mentioned Darwin's important book on the earthworm and the misleading theories of the German scientist Liebig.

Readers may have heard of Sir Robert McCarrison, the doctor of medicine who, in the first quarter of this century, worked up in the north-west frontier, where he discovered a tribe called the Hunzas. This tribe was particularly free from diseases – from cancer, duodenal ulcers, gastric ulcers and the like. The only difference between the Hunza and the other tribes on either side of them was that they were believers in putting back into the soil that which was taken out. The Hunzas were great composters.

In a lecture to the Royal Society in Great Britain, Sir Robert McCarrison said:

There is something in the freshness and quality of food, which is not accounted for by the known chemical ingredients of food – proteins, fats, carbohydrates, minerals and vitamins. It is certain that this something plays a part in that perfect combination of the eye, the muscle, the nerve, the blood vessels, the endocrines, which enable the heron to avoid the hawk, and in their other protective equipment, serenity and courage, clean blood plasma and a lively system, if I may call it that, which wards off infection and constitutes a natural active immunity.

This means, of course, that if somebody's perfectly healthy in body, they're more likely to be able to resist the attacks of disease. There is no doubt that there's a very definite correlation between the food one eats and health. Some years ago *The Medical Testament* by D. L. J. Picton, a doctor in Cheshire, England, was published. The ex-

131

perience of Dr Picton and his large team of research scientists was that it was impossible to separate human health from the health of the animal to be used as food. These doctors said that whether it was a fruit or a vegetable, or animals feeding on herbage, all of them undoubtedly had their origin in perfect soil fertility. It is essential to have what I call a live soil; but to have a live soil, there must be plenty of humus present.

Humus is a kind of jelly, or colloid, which can stick together the particles of a sandy soil and make them better for growing crops. This can keep apart the very tiny particles of a clay soil and so make a clay soil more open. Lots of people who come to see my garden say, 'Of course, anybody can adopt this method in your soil, because it's a first-class soil.' I have to tell them that my soil is terribly difficult. It is the most horrible London clay – yellow and grey and sticky – with about 150 mm (6 in) of gravel above it; and this is one of the worst soils in which to grow flowers or food. But because the people who came to Arkley Manor garden see the layer of compost on top of the earth they think they're seeing the soil. The worms will have pulled in what they need, the bacteria will have worked on it and produced the plant foods, but lying on the top, always, is this layer, because if the worms do pull in a fair amount we top it up, and thus the 25 mm (1 in) thick layer is there all the time.

Now, some people have called this business of humus in the soil a protective fence. Twenty-five years ago, Professor Maisin of the Cancer Institute in Louvain, Belgium, wrote: 'In certain vegetables and foodstuffs, there are one or two factors which are active in the prevention of cancer, caused by the azo-dyes and hydro-carbons.' Cancer is one of today's real dangers, with which, I think, all of us must be concerned. A few years later, the great cancer specialist Bauer (a surgeon, be it noted, not a nutritionist) confirmed this statement with the words: 'We are justified in concluding that nutrition ... can favour cancer, or, on the other hand, protect against it.'

Excellent results in the control of the chrysanthemum eel-worm (to go on to another disease) have been achieved by the use of compost in Great Britain. The special beneficial fungi living in the compost actually surround the eel-worms and suck them to death as they feed on them.

Dr R. V. Harris, the great mycologist late of the East Malling Research Station, Kent, has said:

> There are places where strawberries succumb to virus. There are other places where they flourish, even when virus-infected plants are established close by. In the 1930s, Hampshire strawberry fields suffered a bad slump, and in a desperate effort to cut costs and increase production, strawberry-growers resorted to chemical fertilizers on a heavy scale; and it was about this time that the ravages of virus made themselves felt, and became progressively worse ... Some say there's no connection between these two phenomena. On the other hand, I have recently visited some very fine strawberry plantations in Hampshire, free from all traces of virus, and in every case the soil has been built up to a high state of fertility with organic manures only.

I want you to note the word 'only'.

In the middle of the seventeenth century a certain Dr van Helmont planted a 2 kg (5 lb) willow tree in 90 kg (200 lb) of dry earth. He added nothing but rainwater to the soil for five whole years and the tree grew to weigh more than 76 kg (169 lb) while the earth, dried and re-weighed, lost only 56 g (2 oz). His conclusion was that plants were formed solely from water. The doctor didn't realize, of course, that the tree had absorbed carbon and oxygen from the air, and that the living organisms in the soil made much nitrogen available, and further that the water itself, in addition to supplying hydrogen and oxygen, had carried up with it some plant foods.

The ninety-two elements in the earth and its atmosphere are all naturally occurring substances that are not decomposed during chemical change. The maintenance of health and natural growth are certainly associated with at least thirty-seven of the almost sixty elements found in either plants or animals. I was assured when I went to lecture in the United States that it had been shown that the better the land and the better the farming practices, the better the men! Good health is certainly more the rule and less the exception when the right quantities of the essential elements are present in the soil. Carbon, nitrogen, oxygen and hydrogen make up approximately 95 per cent of the dry weight of all plants; and calcium, phosphorus and potassium compose most of the remainder. Some are supplied by the rain as it falls or are naturally in the soil. Others of course are supplied when properly made compost is added, either as a top dressing or by lightly rotivating it in.

Because the micro-nutrient minerals are required in such tiny quantities, some have felt that they are not as necessary as those needed in bulk. Remember that they account for only 1 per cent of the total weight of plants! It is however this often-overlooked 1 per cent that can make all the difference between a really healthy crop rich in vitamins and a poor crop of little food value.

Elements that are essential for life are *all* equally important, though the quantities required of one may be very small indeed in relation to another. Plants for instance need only very small traces of zinc; on the other hand they need large amounts of available phosphorus in the soil. Yet I have discovered that a deficiency of *either* element can ruin what could have been an otherwise good crop.

The essential minerals for plant life can be divided into two groups. Calcium, phosphorus, and sulphur, together with sodium, potassium, chlorine and magnesium, should be classified as the macro- or 'big' nutrient minerals. It must be realized however that this high-sounding title has nothing to do with their importance. It is simply used because these elements are needed in larger amounts by the plants than are what are called the micro-nutrient minerals or trace minerals, which are today, all over the world, receiving more and more attention. They include iron, copper, zinc and manganese, essential for plant *and* animal life, together with boron and molybdenum, necessary for plants, and iodine and cobalt, indispensable for animal life.

These micro-nutrients are required in minute quantities. The human body contains only about 3 parts per million of manganese, but plants contain only about one-hundredth of 1 per cent! Zinc occurs in plants in amounts much smaller, ie about 50 parts per million. How these trace elements operate is a mystery. Iron is an esential part of the blood, just as copper is of all green plants. Cobalt is present in Vitamin B_{12}, the anti-pernicious-anaemia factor, while boron is fundamental if good new plant roots are to develop properly. Molybdenum plays a role in the metabolism of nitrogen in plants. Copper is regarded as a catalyst in the formation of blood, while manganese is a part of the enzyme that enables carbon to be used by plants.

One of the difficulties connected with the use of the micro-nutrients is the narrow range that invariably exists between correct soil concentrations that will do away with deficiency symptoms and those that will cause harm and toxicity. An example is found in the case of boron. Some plants show boron deficiencies when the soil solution contains less than 0.01 of a part per million. These amounts are extremely small and impossible to measure by the gardener.

None of the micro-nutrients can, however, act alone. There is, I believe, no such thing as a one-element deficiency. All elements seem to fit into the life-and-death pattern and have a very definite purpose, working together. As scientists continue to explain the functions of the various elements, gardeners will be better able to use them more intelligently. In the meantime, their usage must be by the application of correctly made compost.

It is now well known that iodine is a protection against goitre. Before iodine was regularly added to grocer's salt, goitre was very common among the people who lived in the Great Lakes region of America. It was characterized by an enlargement of the thyroid gland. Plants, of course, absorb it and pass it on. In fact, iodine is added naturally to the soil through seaweed manure. The result is that the people who eat plants grown on soil manured in this way never suffer from goitre.

Cobalt is necessary to the health of animals. In Australia where I have lectured year after year, the lack of this micro-nutrient results in the coast disease, bush sickness in humans, and 'the skinnies' in animals. When cobalt is lacking, animals lose their appetites and their condition.

The opinion of Dr William Albrecht of the University of Missouri confirms that soil is truly the basis of nutrition of microbes, plants, animals and therefore man, at the top of what he calls the biological pyramid. The better our understanding of humus and all micro-nutrient minerals the better our gardening techniques.

What influence is all this knowledge likely to have in maintaining your soil in a fertile condition, and the health of you and your plants? Is the adding of humus to the soil one of the keys to the successful fighting of disease? Is the answer that we shouldn't wait to combat the serious and painful symptoms brought about by disease, but obviate the cause? I think that there is much work for the research worker to do on this vast subject. But I think, too, that the gardener is called upon now more than ever to make his practical contribution on the lines I've laid down. I have, at any rate, tried to give a balanced and unbiased account on this subject to date, and I can only hope and pray that this will provide a talking-point and much thought.

INDEX

137

Mayflower Handbooks for your information

Cooking and Drink

ON DRINK	Kingsley Amis	60p	☐
PICK, COOK AND BREW	Suzanne Beedell	75p	☐
SUCCESSFUL MODERN WINE-MAKING			
	H. E. Bravery	40p	☐
GREEK COOKING	Robin Howe	75p	☐
GERMAN COOKING	Robin Howe	75p	☐
COOKING THE CHINESE WAY	Kenneth Lo	50p	☐
THE SCOTS KITCHEN	F. Marian McNeil	75p	☐
BEE NILSON'S SLIMMING			
COOKBOOK	Bee Nilson	95p	☐
MAKING ICE-CREAM AND COLD SWEETS			
	Bee Nilson	75p	☐
DEEP FREEZE COOKING	Bee Nilson	60p	☐
FONDUE, FLAMBÉ AND SIDE TABLE			
COOKING	Bee Nilson	60p	☐
BEE NILSON'S KITCHEN HANDBOOK			
	Bee Nilson	50p	☐
CAKES AND PUDDINGS	Bee Nilson	50p	☐
BEE'S BLENDER BOOK	Bee Nilson	60p	☐
MEATS AND ACCOMPANIMENTS	Bee Nilson	40p	☐
BEE NILSON'S BOOK OF LIGHT MEALS			
	Bee Nilson	40p	☐
INDIAN COOKERY	E. P. Veeraswamy	60p	☐
VEGETARIAN COOKERY	Janet Walker	50p	☐
LET'S PRESERVE IT	Beryl Wood	60p	☐

Gardening

PEARS ENCYCLOPAEDIA OF GARDENING:			
FRUIT AND VEGETABLES	ed Roy Genders	£1.25	☐
PEARS ENCYCLOPAEDIA OF GARDENING:			
FLOWERS, TREES AND SHRUBS			
	ed Roy Genders	£2.50	☐

Sport and Health

YOGA MADE EASY	Desmond Dunne	60p	☐
HOW TO PLAY WINNING TENNIS	Rod Laver	35p	☐
NATURAL REMEDIES FOR COMMON			
AILMENTS	Constance Mellor	75p	☐
YOGA AND YOUR HEALTH	Sonya Richmond	50p	☐
SLIMMING: AN ORIENTAL APPROACH	Soraya	40p	☐
KARATE	Bruce Tegner	50p	☐

Mayflower Historical Fiction for your enjoyment

Favourite reading from Mayflower Books
A list of established bestsellers that you may have missed

FANNY HILL	John Cleland	60p	☐
THE EGG AND I	Betty Macdonald	40p	☐
THE STORY OF SAN MICHELE	Axel Munthe	75p	☐
THE FOUNTAINHEAD	Ayn Rand	95p	☐
THE GADFLY	E. L. Voynich	40p	☐
THOSE ABOUT TO DIE	Daniel P. Mannix	50p	☐
THE HINDENBURG	Michael M. Mooney	60p	☐
GET ME TO THE WAKE ON TIME			
	ed Alfred Hitchcock	35p	☐
DEATH BAG	ed Alfred Hitchcock	40p	☐
BAR THE DOORS	ed Alfred Hitchcock	35p	☐
A HANGMAN'S DOZEN	ed Alfred Hitchcock	35p	☐
THE STORIES OF FLYING OFFICER 'X'			
	H. E. Bates	50p	☐
PATROL	Fred Majdalany	30p	☐
THE BATTLE OF CASSINO	Fred Majdalany	75p	☐
ALL QUIET ON THE WESTERN FRONT			
	Erich Maria Remarque	50p	☐
AS FAR AS MY FEET WILL CARRY ME			
	J. M. Bauer	50p	☐
HUNTING THE BISMARCK	C. S. Forester	25p	☐
LONELY WARRIOR	Jean Offenberg	40p	☐
POPSKI'S PRIVATE ARMY	Vladimir Peniakoff	75p	☐
IMPOSSIBLE POSSIBILITIES	Pauwels and Bergier	50p	☐
ETERNAL MAN	Pauwels and Bergier	50p	☐
THE MORNING OF THE MAGICIANS			
	Pauwels and Bergier	75p	☐
THE GREAT BEAST	John Symonds	60p	☐
THE KANDY-KOLORED TANGERINE-FLAKE			
STREAMLINE BABY	Tom Wolfe	40p	☐
UNIDENTIFIED FLYING OBJECTS			
	Robert Chapman	50p	☐

Mayflower Personality Books for your enjoyment